ORIGINS OF THE FRENCH REVOLUTION

ORIGINS OF THE FRENCH REVOLUTION

by
WILLIAM DOYLE

Second Edition

OXFORD UNIVERSITY PRESS

Oxford University Press, Walton Street, Oxford OX2 6DP
Oxford New York Toronto
Delhi Bombay Calcutta Madras Karachi
Petaling Jaya Singapore Hong Kong Tokyo
Nairobi Dar es Salaam Cape Town
Melbourne Auckland
and associated companies in
Berlin Ibadan

Oxford is a trade mark of Oxford University Press

Published in the United States by
Oxford University Press, New York

First published 1980
Reprinted 1981
Reprinted (with corrections) 1982, 1984, 1985, 1987

Second edition, 1988
Paperback reprinted 1989

British Library Cataloguing in Publication Data
Doyle, William
Origins of the French Revolution.
1. France—History—Revolution, 1789–1799—
Historiography
2. France—History—Revolution, 1789–1799—
Causes and character
I. Title
944.04'072 DC147.8 80-40740
ISBN 0-19-822283 1
0-19-822284 x (Pbk)

Printed in the United States of America

To Frances and Ed
after ten years

Prefatory Note

Half a generation of students have helped me to clarify my ideas about this subject—in so far as I have. Colleagues and friends too numerous to mention have also played their part. I am particularly grateful to Norman Hampson, who has read through the whole text with characteristic care and critical detachment, offering countless suggestions for its improvement. Ted Royle was also kind enough to read a crucial chapter and give me his comments at a very busy time. Christine, as always, was relentless in her criticism and unstinting in her help.

September 1979 *W.D.*

Preface to the Second Edition

In the decade since the manuscript of this book was completed much important new work has appeared on many of the areas with which it deals. Very little of this new work, however, seems to me to challenge the basic lines of argument which I have put forward; so, in revising the text I have only taken later scholarship into account when it has shown that I made factual errors, or when it has fundamentally affected leading aspects of my argument. This is certainly not to disparage the labours of those who have carried the study of the subject forward since the late 1970s. But to give them all the credit they deserve would require a new book rather than a mere second edition of an old one.

November 1987 *W.D.*

Contents

Les origines de la Révolution sont une histoire; l'histoire de la Révolution en est une autre.

D. Mornet
Les Origines intellectuelles de la Révolution française

Introduction

Why was there a French Revolution? This is a question of perennial interest not only to scholars, but to everybody who takes an interest in the history of the world. Naturally, therefore, it is also a subject of endless controversy. And yet towards the middle of the twentieth century something like a consensus seemed to be emerging, among scholars at least, and disagreement was confined to small corners of the subject which did not materially affect the general picture. This consensus, very properly it seemed, was most clearly expressed by the French masters of the period, by Albert Mathiez, by Georges Lefebvre, and by Ernest Labrousse. English-speaking historians largely confined themselves to a respectful rehearsal of Frenchmen's own version of their history, and younger French scholars certainly showed little enough inclination to challenge what had solidified into an orthodoxy.

Yet since the 1950s all this has changed. Under the onslaught of new research, mainly conducted by non-French scholars, the consensus began to crumble, and in 1962 it was subjected by Alfred Cobban to a frontal attack. After that no part of the old view remained sacrosanct, and by 1970 the occasional sceptical voice was being raised in France itself. Scores of American, British, and other foreign research students invaded the French archives and later proclaimed their dissent—for their own corners of the field—from the old orthodoxies. The flow of their findings has by no means stopped, and it has again made the origins of the French Revolution one of the more lively areas of historiography.

So the consensus has collapsed, and controversy once more rages. The difficulty is to see what, if anything, the masses of new material and reinterpretation that have appeared in the last thirty or so years amount to. They have destroyed one edifice: can a new one be built from the debris? Or must the question of the Revolution's causes continue, as the reconstructor of another revolution's origins has complained, to wallow in fragmented chaos?[1] It seems to me that those who have so revelled in the joy of destruction owe it to their fellow scholars, and not least to their students, to declare what they now think the general picture to be. The bewilderment of students faced with a huge mass of complex and often waspish articles, rather than any clear outline of the problem, must be familiar to everybody who has attempted to teach this subject. Too many of them, daunted by the reading required to come abreast of the most recent scholarly opinion, prefer to take refuge in the old cliches, as if a generation of new research had never been accomplished. Nobody can feel satisfied with such a situation.

What follows is not an attempt to close the case on the origins of the French Revolution—an impossible task in any event. It is rather an attempt to summarize and explain progress so far, to see where we have got since the safe and certain days of Mathiez and Lefebvre. It is one man's attempt to state clearly and concisely how recent research has altered our view of what caused the greatest of all revolutions, and to incorporate our new knowledge into a coherent and credible explanation. The essay falls naturally into two parts. In the first, I have tried to summarize the developments in historiography and research since 1939 which have produced the present 'state of the question'. Then I have tried to take account of all these developments in constructing a new analysis of the breakdown of the old regime, the struggle for power which it brought about, and the reasons why these developments occurred as they did. The two parts are meant to be complementary, but there is no reason why they should not be read in isolation from each other. So those more interested in what I think than in what others have thought can bypass part I. Those interested in why I think it will find my sources in the footnotes. For reasons of economy I have omitted a bibliography at the end; by then most works of importance will have been cited already.

Nobody could undertake a work like this in any confidence of winning general agreement with his conclusions. But if I can persuade my readers to start disagreeing with me rather than with old orthodoxies that are not only dead but now in urgent need of burial, then perhaps the debate will have made a permanent advance.

PART I

A Consensus and Its Collapse: Writings on Revolutionary Origins since 1939

The year 1939, which saw the outbreak of the Second World War, is also remembered among students of the French Revolution as the 150th anniversary of the storming of the Bastille. In France the occasion was marked by extensive celebrations and a spate of new writings on the Revolution and its significance. Undoubtedly the most distinguished product of this activity was Georges Lefebvre's concise account of the origins of the Revolution and its outbreak, *Quatre-Vingt-Neuf.* Unashamedly exulting in the achievements of 1789, Lefebvre's book itself had an eventful history in the years following its publication. Much of the edition was destroyed, as subversive literature, on the orders of the Vichy government. In 1947, however, it was translated into English by the American scholar Robert R. Palmer; and under the title of *The Coming of the French Revolution* it rapidly became essential reading wherever the French Revolution was studied throughout the English-speaking world. By the time of Lefebvre's death in 1959 it had sold over 40,000 copies in English. It remains, and will remain, a classic: a model of historical writing by a master of his subject. For many years its analysis of the death throes of the old order in France, and the diseases which killed it, remained definitive.

The ultimate cause of the French Revolution, Lefebvre believed, was the rise of the bourgeoisie.[1] He was a Marxist, and Marxism is a theory of history which assigns a central role to the bourgeoisie as the representatives and beneficiaries of capitalism.[2] According to Lefebvre, 1789 was the moment when this class took power in France, after several centuries of growing numbers and wealth. Medieval society had been dominated and ruled by a landed aristocracy, for land was the only form of wealth. By the eighteenth century, however, 'economic power, personal abilities and confidence in the future had passed largely to the bourgeoisie', who were buttressed by 'a new form of wealth, mobile or commercial' and a 'new ideology which the "philosophers" and "economists" of the time had simply put into definite form'. In 1789, the bourgeoisie overthrew the remnants of the aristocratic, landed order which had till then retained social predominance despite its economic eclipse, and established a regime which more closely reflected the new distribution of economic power. Yet there was nothing automatic in this development, at least in the short run. The bourgeoisie were able to overthrow the aristocracy because the political authority of the monarchy had collapsed. It had collapsed because the monarchy was unable to pay its way. And it was unable to pay its way because the aristocracy, the 'privileged orders' of nobility and clergy, clung to their exemptions and privileges, and used their political power to prevent the king from making the necessary reforms.

Between 1787 and 1789, however, Lefebvre discerned not one revolutionary movement but four. First came the revolt of the aristocracy, which destroyed the monarchy, and was the climax of a century-long aristocratic resurgence or reaction, in which the nobility had sought to regain the preeminence in the state of which Louis XIV had deprived them. In order to effect their revolution, the nobility had recruited the support of the bourgeoisie, but the very success of the movement gave the bourgeoisie ideas of their own. When in September 1788, the parlement of Paris, the very spearhead of the aristocratic reaction, declared that the Estates-General promised by the government for 1789 should be constituted as they had been in 1614 (their last meeting), the bourgeoisie exploded with fury; for the 'forms of 1614' guaranteed aristocratic predominance. Thus began the revolution of the bourgeoisie, a class struggle against the aristocracy which lasted until the creation of the bourgeois-dominated National Assembly in June 1789. The bourgeois aim was civil equality; they wished to destroy the privileges of the nobility and the clergy and establish a regime where all men obeyed the same laws, paid taxes on the same basis, enjoyed the same career opportunities, and owned property on the same terms. These ideals stemmed from the Enlightenment, the intellectual product of the rise of the bourgeoisie. The collapse of the monarchy in 1788 gave the bourgeoisie an opportunity of which they had long been dreaming to put them into practice. But, just as the aristocracy had marshalled bourgeois support to defeat the monarchy, so the bourgeoisie needed other elements in order to consolidate their own success. In July 1789, their precarious victory was threatened by a noble-inspired royal attempt to dissolve the National Assembly. The *coup* was only defeated by an uprising of the Parisian populace, whose most spectacular achievement was the storming of the Bastille. This third revolution, the popular revolution, was a movement which sprang from hope that the new order would resolve the growing economic problems of urban workers. These same workers intervened decisively again in October 1789, when the Estates-General once more seemed threatened by a royal-aristocratic counter-*coup*. Before that happened, however, the economic crisis of 1788–9 had produced a fourth revolution, the peasant revolution—a nation-wide uprising, fired by panic fears for the safety of ripening crops—against the exaction of seignorial dues and labour services by aristocratic landlords. This movement was only stilled by the abolition, on the night of 4 August 1789, of the whole apparatus of this 'feudalism', the last bastion of the old aristocratic order.

Such a brief summary inevitably distorts a work as subtle and skilfully written as *Quatre-Vingt-Neuf.* To appreciate its full flavour there is no substitute for reading the book itself. In addition, Lefebvre's argument can be sampled in the general survey of the 1790s which he wrote twelve years later.[3] The interesting problem is why his interpretation of the origins of the Revolution was able to command such widespread support. After all, in the 150 years since it had occurred, few historical events had aroused more sustained and impassioned controversy.

One reason was that, by 1939, Lefebvre was the undisputed master of the subject. Professor of the History of the French Revolution at the Sorbonne, he had built up a formidable scholarly reputation through a series of monographs which had transformed understanding of the agrarian history of the Revolution.[4] His exact contemporary, and only possible rival, Albert Mathiez, had died prematurely in 1932. With him died a bitter, polemical tradition in revolutionary historiography which Lefebvre, for all his commitment to many of the ideals and interpretations of Mathiez, avoided. Another strength of Lefebvre's interpretation was that it took full account of the researches of Ernest Labrousse into the economic origins of the Revolution.[5]

Labrousse's analysis of price movements down to 1789 gave economic substance to the discontent expressed by all classes in 1789, and lent impressive strength to the argument of Lefebvre. The publication of Labrousse's next study in 1944, which described in particular detail the crisis in wine production between 1778 and 1789, reinforced rather than modified Lefebvre's explanation of peasant discontents.

The most bitter debates about the Revolution in the earlier part of the century had had strong political overtones. Historians of the left, like Jean Jaurès and Mathiez, believed that the Revolution was both desirable and inevitable. Those of the right, like Pierre Gaxotte, Bernard Faÿ, and Frédéric Braesch, saw it as abominable and something which, if the proper steps had only been taken, might, and should have been avoided.[6] The basic disagreement was about democracy, democratic principles, and public order; and nobody could doubt where Lefebvre stood on this. He was a man of the left, a believer in popular power, and he thought that the Revolution of 1789 could not have been effected without violence. The Vichy regime which suppressed his book was well aware of the fact. And yet he propounded a view of what caused the Revolution that even the most conservative of French historians could broadly accept. For the quarrel between left- and right-wing historians was largely about the Revolution's consequences. On the question of its origins, the two sides did not fundamentally disagree at all. Both acknowledged that the rise of the bourgeoisie was the fundamental cause. Both agreed that the monarchy tried to reform itself, and that it was prevented from doing so by the selfish obstruction of the aristocracy. Both believed that the Enlightenment had sapped faith in traditional values (irrespective of whether they welcomed or deplored the fact). In these circumstances Lefebvre's interpretation passed largely unassailed by what might have seemed potentially its most vehement critics.

In fact, *Quatre-Vingt-Neuf* passed unassailed by almost anyone, thanks largely to the Second World War. Not only were most copies destroyed; scholarly activity in general came to a halt, and very few reviews appeared before that happened.[7] Even more important, hardly any new research was done during the war or for several years after it ended. Most work published in the 1940s was based upon research done during the previous decades.[8] All these circumstances combined to leave Lefebvre's account of the Revolution's origins unchallenged for the best part of twenty years after it appeared. The first major post-war history, a subtle, brilliant, and concise account by the Englishman Albert Goodwin, contained an analysis of the movement's origins that differed hardly at all from Lefebvre's, and plainly owed a good deal to him.[9] Almost a decade later, the first general work on the Revolution from a new generation of French historians,[10] by Albert Soboul, added nothing to Lefebvre's account except applause, and a more explicit bias towards Marxism than the latter had ever shown. Only in 1965 did François Furet and Denis Richet presume to challenge what had become a rigid orthodoxy in France; and even then their dissent was over the course of the Revolution rather than its origins.[11]

Outside the narrow, largely monolingual world of French scholarship, however, doubts had begun to arise much earlier —and fundamental doubts, too. Alfred Cobban, newly appointed Professor of French History at the University of London, chose the occasion of his inaugural lecture in 1954 to attack what he called the 'Myth of the French Revolution'.[12] This myth, which he contended had dominated serious research on the history of the Revolution during the twentieth century, was that the Revolution was 'the substitution of a capitalist bourgeois order for feudalism.'[13] Cobban argued that anything that could conceivably be called feudal had passed away long before 1789, and that to use the term for the complex of rights and dues that were swept away on the night of 4 August 1789 (as had Lefebvre and other writers before him) was to confuse rather than clarify matters. Gleefully, Cobban picked out Lefebvre's own admission that many bourgeois in the National Assembly owned such rights and dues and were reluctant to relinquish them. This would have been awkward to explain if the revolutionary bourgeoisie had been, as Lefebvre contended, the representatives of mobile, industrial, and commercial wealth —capitalists, in a word. But Cobban argued that they were not. By analysing the social and professional background of the bourgeois elected to the Estates-General in 1789, he showed that only 13 per cent of them came from the world of commerce, whereas about two-thirds were lawyers of one sort or another. Goodwin had already dwelt on this,[14] but Cobban carried the analysis much further. Forty-three per cent of the bourgeois deputies were not only lawyers, he pointed out, but petty office-holders and government servants. And he went on to argue that it was the frustrations of such people, doing the work of government but receiving none of the credit, and seeing the value of the venal offices in which they had invested their money declining, which provided the reforming impetus of the revolutionary bourgeoisie, and its commitment to careers open to the talents. From this major conclusions followed: the Revolution was not the work of a rising bourgeoisie at all, but rather of a declining one. The revolutionaries of 1789 were not hostile to 'feudalism', which the peasants, not they, destroyed; nor were they the standard-bearers of capitalism. Cobban incorporated these views into his own general survey of eighteenth-century France, published two years later.[15]

At first they did not make the impression they deserved. Inaugural lectures are seldom widely read, and to bury new ideas in a general survey is to risk many readers overlooking their importance. Besides, Cobban's revisionism was very rapidly dismissed by Lefebvre himself. Writing in the one journal read by all serious students of the Revolution, the *Annales historiques de la Révolution française*, which he himself edited, he declared that Cobban was trying to deprive the Revolution of its fundamental significance.[16] Cobban's strictures on the use of the term feudalism were a quibble about terminology, and although his analysis of the membership of the revolutionary assemblies was of value and interest, the conclusions he drew missed the point. Even if the men of 1789 were not capitalists, their actions favoured capitalism's future development, and that was the most significant thing. To suggest anything else was to denigrate the Revolution, and Lefebvre suggested that this was Cobban's true aim. As a representative of the twentieth century's ruling orders, the Englishman was frightened of revolutions, and sought to minimize the inspiring precedent and example of 1789. The 'myth', if myth it was, was a necessary one.

But Cobban was not so easily disposed of, nor were the problems he raised. As so often in historical research, he had merely been the first to express views towards which other scholars were also working quite independently. As early as 1951, an American doctoral student, George V. Taylor, had argued in an unpublished thesis that those members of the bourgeoisie who were capitalists were largely uninterested in politics both before and during the Revolution, except as a possible vehicle for protecting their own commercial and industrial privileges.[17] The implication was that they did not share the aspirations of the reformers in the Assembly, and that therefore the bourgeoisie was far from the united, self-conscious and self-confident class that had often been assumed. The only important French historian besides Lefebvre to deign to notice Cobban's arguments in the late 1950s, Marcel Reinhard, also noted that commercial and professional members of the bourgeoisie were often deeply suspicious of one another; he called for more research into such matters.[18] The few pieces of archival research into the bourgeoisie that now began to appear also concluded with the need for more research; but in their own limited fields of enquiry they emphasized the diversity of the eighteenth-century bourgeoisie, and the wide range and varied composition of its fortunes.[19] Only works based on literary sources rather than archival research continued to be more preoccupied with what bound the bourgeoisie together than with its inner divisions and contradictions.[20] In 1964, therefore, when Cobban returned to the attack, the atmosphere was more receptive—and Lefebvre, who had died in 1959, was no longer there to concert a defence of the accepted wisdom.

In *The Social Interpretation of the French Revolution,*[21] the published version of a series of lectures delivered in 1962, Cobban argued his case of 1955 in more detail. He now defined 'feudalism' as seignorial rights and dues, but continued to argue that it was the peasants rather than the bourgeoisie who really opposed the system and overthrew it. Above all, he returned to his theme that 'the revolutionary bourgeoisie was primarily the declining class of *officiers* and the lawyers and other professional men, and not the businessmen of commerce and industry.'[22] Far from promoting capitalism, the Revolution was to retard it; and that was, indeed, the fundamental aim of the various revolutionary groups in 1789. 'In so far as capitalist economic developments were at issue, it was a revolution not for, but against, capitalism. This would, I believe, have been recognised long ago if it had not been for the influence of an un-historical sociological theory. The misunderstanding was facilitated by the ambiguities implicit in the idea of the bourgeoisie. The bourgeois of the theory are a class of capitalists, industrial entrepreneurs and financiers of big business; those of the French Revolution were landowners, *rentiers*, and officials.'[23] He added that the latter group were bitterly at odds with the former during the closing decades of the old order. And what gave special piquancy to Cobban's onslaught on orthodoxy was the way he used the arguments and researches of its greatest champions, such as Lefebvre himself and Soboul, to support his own case and undermine theirs. How could Soboul, he asked, declare dogmatically that the triumph of the bourgeoisie was the essential fact of the Revolution, when in the same breath he admitted that the history of the bourgeoisie during the Revolution remained unwritten?[24] The final chapter of Cobban's book was an analysis of Lefebvre's last work,[25] which had just been posthumously published, in which he showed that Lefebvre's findings on the social structure of the pre-revolutionary Orléanais lent less support to the general views of their French author than to those of his English critic.

This time Cobban's opinion made some impact in France, if only to invite indignant refutation from the established academic authorities. A totally hostile review by Jacques Godechot—then, in terms of volume of published work at least, the leading French authority on the Revolution—provoked a series of spirited and entertaining replies from Cobban.[26] Even English-speaking reviewers had their breath taken away by the scale and range of Cobban's attack. Unlike the French, most of them seemed inclined to accept the destructive side of his argument, but they were deeply sceptical of what he proposed to put in place of the orthodoxies he had undermined. Most reviewers seemed to agree (for a wide range of reasons) that Cobban's arguments raised far more questions than they answered. The appearance of his book, in fact, signalled the beginning of a period of acute controversy in the historiography of the Revolution, not unlike the 'storm over the gentry' which had already been raging for almost two decades among historians of the English revolution of the seventeenth century.[27]

The credit (or blame) for this did not lie with Cobban alone. By the mid-1960s it was evident that old views were crumbling in every direction. Norman Hampson, for example, who stigmatized Cobban for having produced nothing more than a 'non-Marxist economic interpretation of the revolution',[28] had already laid stress, in a new general survey published in 1963, on the ambiguities of the term bourgeoisie. He emphasized that social and economic motivations were not the same thing, and he stressed the importance of intellectual convictions in bourgeois hostility to the old order.[29] Cobban had, by contrast, always tended to belittle the role of the Enlightenment in the origins of the Revolution—another way in which he differed from Lefebvre.[30] Hampson still, however, saw a fundamental opposition between the 'privileged orders' and their interests on one side, and the bourgeoisie and its interests on the other. Neither Cobban nor his French opponents would have disagreed. But the great development of the late 1960s was to call even this into doubt.

Nothing contributed more to a reappraisal of the revolutionary bourgeoisie than new work on its supposed antithesis, the nobility. Yet this was a subject on which Cobban was completely orthodox. He hardly ever expressed a view about the nobility with which Lefebvre would have wished to disagree. He believed that the nobility was a selfish, increasingly exclusive caste which used its political power, wielded through the parlements, to prevent the crown from attacking its privileges. Eventually noble obstruction brought the old order down.[31] Unlike the traditional picture of the bourgeoisie, that of the nobility was to be slowly eroded rather than shattered by frontal attack.

The drift of two future decades of scholarly opinion was, indeed, foreshadowed in 1953 by John McManners. In an impressionistic essay buried in a collection devoted to European nobilities in the eighteenth century, he argued that money, not privilege, was the key to pre-revolutionary society in France. Wealth transcended all social barriers and bound great nobles and upper bourgeois together into 'an upper class unified by money.'[32] This essay was not a work of original research, but the previous year Jean Egret had used new French material to argue that the exclusivism of the parlements, the supposed spearhead of the 'aristocratic reaction' which preceded the Revolution, had been much exaggerated.[33] Such work suggested that the rigidity of the pre-revolutionary social structure had been overdrawn, and deserved further research; but as yet this was not forthcoming. The most influential work of the early 1950s, in fact, merely lent new arguments to established opinions. This was Franklin Ford's *Robe and Sword,*[34] which re-examined the customary division of the nobility into nobles of the sword and nobles of the robe. The former, according to tradition, were of old stock; they looked down upon the latter's recent ennoblement by judicial office. Ford now argued that by the early eighteenth century many of the robe nobility no longer owed their ennoblement to their offices, and that by 1748 they had captured the leadership of the nobility as a whole. The old antagonism between sword and robe was largely forgotten as nobles united against commoners on one side and the crown on the other. Ford's book was widely acclaimed as a brilliant contribution to the understanding of eighteenth-century France, for it offered a convincing explanation of why the aristocratic reaction—which most authorities agreed took place in the later eighteenth century—occurred, and why it was such a powerful movement.

The first widely noticed blows to the traditional picture of the nobility did not fall until the early 1960s. In 1960, Robert Forster published a social and economic study of the nobility of Toulouse which showed that they were neither impoverished rustics nor debt-laden prodigals, but shrewd and careful managers of their estates. The key to their social and economic position was to be found in 'adherence to the so-called bourgeois virtues of thrift, discipline and strict management of the family fortune.'[35] A series of articles by the same author over the next few years[36] emphasized the lesson for other areas of France. They suggested that in economic outlook the nobility and the bourgeoisie had much in common. Meanwhile, Betty Behrens had argued that the fiscal privileges of the nobility were not as great as the revolutionaries, and historians following them, had claimed. The French nobility, in fact, was perhaps the most highly taxed in Europe; and the really important tax-exemptions were those enjoyed by the commercial bourgeoisie of the great mercantile cities.[37] Five years later, she expanded on these views to argue that no social group in pre-revolutionary France had a monopoly of privilege. Most of society was privileged in some way or other, and among the most valuable privileges were those belonging to the bourgeoisie.[38] By this time, too, George V. Taylor was beginning to publish his long-awaited findings. In 1964, he produced a careful analysis of the types of capitalism that existed in France before the Revolution, concluding that they bore little resemblance to the industrial capitalism of the future. Above all, eighteenth-century capitalism was far from a bourgeois monopoly. One of its basic features was the heavy involvement of nobles.[39]

The effect of all this work, in demolishing, or at least challenging, the traditional picture of the pre-revolutionary nobility, was to reveal them as more and more like the bourgeoisie; and Taylor did not fail to see the implication. In 1967, he suggested, in an article second only to Cobban's book in its impact, that the wealth of all social groups in pre-revolutionary France was overwhelmingly non-capitalist in nature. 'Proprietary', he suggested, would be a better word.[40] Capitalism had not, therefore, become the dominant mode of production in the French economy before 1789. Moreover, 'even in the well-to-do Third Estate proprietary wealth substantially outweighed commercial and industrial capital. . . .there was, between most of the nobility and the proprietary sectors of the middle classes, a continuity of investment forms and socio-economic values that made them, economically, a single group.'[41]

Discussion in France, meanwhile, was proceeding on parallel lines. Roland Mousnier, a historian of the seventeenth century, had been provoked by Marxist analyses of popular uprisings during that century into formulating a new model of old regime society. Mousnier believed that before the eighteenth century society could not be separated as Marxists analysed it, into classes.[42] A society of classes, he argued, was one where supreme social value was placed upon the production of material goods; class was determined by productive role. But there were also societies of orders or estates, in which social position was determined by social function rather than by relationship to material production.[43] France in the seventeenth century was a society of orders. The highest order was the nobility, whose function was the supreme one of defending the state. The social position of lower orders was determined by their contribution towards supporting the state in lesser ways; thus administrators and magistrates, for example, enjoyed more power and prestige than merchants or tradesmen. Mousnier's explanation of the French Revolution was that, in the course of the eighteenth century, the consensus about social values that underlay the old society of orders began to break down. By the 1750s, production was coming to seem more important than the service of the state, and with the Revolution the new outlook triumphed. A class society replaced the old society of orders.[44]

Whatever we may think of this bold attempt to redirect thinking on social history, it made its mark in France—if only because the power of professors like Mousnier made it advisable for ambitious young academics to pay lip-service to his doctrines. Assertions that pre-revolutionary society was a society of orders began to appear in textbooks and even in works of popularization.[45] By the late 1960s, historians beyond the reach of Mousnier's patronage were also beginning to emphasize what bound the nobility and upper bourgeoisie together into a single plutocratic élite.[46]

The guardians of the old orthodoxies were dismayed to see that denouncing or ignoring sceptical 'Anglo-Saxons' was not enough to preserve their purity, and that their own compatriots were beginning to have doubts.[47] They took refuge in repetition. Soboul, elevated to Lefebvre's chair at the Sorbonne and to the editorship of the *Annales historiques de la Révolution française,* regularly produced surveys of various aspects of the Revolution which dogmatically reiterated the old line. And in 1970, Claude Mazauric roundly attacked the revisionism of Furet and Richet's survey, which had sold well since its publication five years before.[48] Furet and Richet, however, were no longer the isolated young iconoclasts of 1965; they were now among the leaders of an equally well-armed and well-entrenched scholarly phalanx, the *Annales* school of historians. The attention of this group had mainly been engaged by quantitative explorations of early modern history. They had largely left the Revolution alone. Now, however, Furet used the prestigious pages of *Annales* to defend his and Richet's approach, and to revise their account of the Revolution's origins in the light of recent work. In a witty and devastating attack on what he called the 'revolutionary catechism',[49] he denounced Soboul, Mazauric, and their ideological ancestors. Soboul's picture of eighteenth-century society, Furet argued, was a 'neo-Jacobin' one: it accepted the revolutionaries' own myths about the order they had destroyed. In emphasizing antagonisms between the nobility and the bourgeoisie, the 'catechism' ignored all that bound them together, the manifold interests that they had in common but which were increasingly threatened by the demands of a government over which neither had control. Richet had already used the pages of *Annales*, too, to elaborate on a long-term alienation of all property owners from the state.[50] Now at last Cobban received some praise in France for showing how few links the bourgeois revolutionaries of 1789 had with capitalism. Like Taylor, Furet concluded that nobles and bourgeois were economically members of the same class, and that the essence of their wealth was proprietary.

But all such analyses ran into a serious problem. If the nobility and the bourgeoisie had so much in common, why did they become such implacable enemies in 1789? The most radical solution to this problem was suggested by Taylor. His analysis of the nature and distribution of wealth before 1789 implied that the Revolution could not be explained in economic terms, as a clash of opposed interests. Taylor declared himself unconvinced by Cobban's attempt to provide a new economic explanation through the declining bourgeoisie. It was time, he concluded, to revert to a purely political explanation of the Revolution's outbreak. The radical reforms of 1789 were products of a political crisis, and not the outcome of long-maturing social and economic trends. 'It was essentially a political revolution with social consequences and not a social revolution with political consequences.'[51] In a further article, five years later, Taylor underlined this point by demonstrating from the *cahiers* of 1789, the grievance lists drawn up by the electoral assemblies which chose the deputies for the Estates-General, that the mood of France in 1789 was deeply conservative. There was no demand from the grass roots for most of the reforms that were soon to take place—which again emphasized that revolutionary radicalism was a result rather than a cause of the political crisis.[52]

To deny the Revolution any social origins at all, however, has proved too daring a step for most historians. They continue to seek, and to find, trends in the pre-revolutionary social structure that help to explain what happened later. From the mass of important new material that has appeared since the debate on the Revolution's social origins really got under way in the early 1960s, two particularly important conclusions suggest themselves. The first is the impossibility of drawing any clear contrast between the nobility and the bourgeoisie. Nobody has challenged Taylor's contention that the wealth of the bourgeoisie was as overwhelmingly proprietary as was that of the nobility. And whereas nobody denies that the bulk of commercial and industrial wealth was in bourgeois hands, it has been shown that there was extensive noble investment in these fields, too.[53] It has even been argued that if any social group represented a restless, aggressive, innovatory capitalism before 1789, it was the nobility rather than the timid and conservative bourgeoisie.[54] There is equally little sign of the bourgeoisie's outstripping the nobles in wealth, whether in the countryside, in stagnant towns like Bayeux or Toulouse, or in major industrial centres like Lyons.[55] Only in booming seaports like Bordeaux had bourgeois fortunes equalled or surpassed those of the local nobility.[56] And even there, as everywhere else, there was a constant passage of the richest into the ranks of the nobility. This pattern of behaviour had long been noticed, and is constantly emphasized by new research. For instance, between 1726 and 1791, 90 per cent of the Farmers General of taxes, often considered the supreme example of bourgeois men of wealth, were noble.[57] Less noticed until recently was the logical implication of the bourgeoisie's abandoning of their status in order to become noble: the nobility cannot have been a closed caste. And so far from that, one recent estimate suggests that at least a quarter of all noble families in 1789 had been ennobled since the beginning of the century.[58] The basis of these calculations has indeed not gone unchallenged,[59] but the general conclusion remains: the nobility was an open élite, not a hereditary caste apart. Nor is it now possible to maintain that this élite grew less open as the eighteenth century went on, thanks to some exclusivist 'aristocratic reaction'. Most recent evidence suggests that institutions that were largely noble in composition in the 1780s had been so a century earlier, and that apparent attempts to discriminate against non-nobles were really signs of antagonism between different sorts of nobles.[60]

The second major theme to emerge from recent research follows from this. If what divided nobles from bourgeois once received too much attention, what divided bourgeois from bourgeois and noble from noble did not receive enough. Research on various aspects of the bourgeoisie continues to confirm the suggestions of Cobban and Taylor about the lack of contact or sense of common interest between the bourgeoisie of the professions and that of trade.[61] It is also becoming increasingly clear that even among the lawyers and *officiers* there was far from universal agreement about what reforms were desirable in 1789.[62] As to the nobility, earlier historians were always anxious to show how much it deserved its fate after 1789; accordingly, they were happy to repeat the old scornful dismissal of the noble order as a 'cascade of disdain', in which every member despised the next. Yet they were equally willing to portray the nobility in 1788–9 as united and determined in pursuit of power. How, if this were the case, the nobility was so easily overthrown was not a question these historians chose to ask, let alone answer. If, however, attention is concentrated on divisions within the nobility, this problem becomes far less difficult, and new research is increasingly emphasizing such divisions. A massive study of the Breton nobility in the eighteenth century has revealed, for example, what deep antagonism there was in this noble-swamped province between rich and poor nobles, and how hostile were the resident nobility to the gilded absentees of Versailles.[63] The famous Ségur ordinance of 1781 has been convincingly shown to be directed not against bourgeois aspirants to military commissions, but against new nobles and their offspring.[64] Despite Franklin Ford's famous argument, robe and sword nobles were still attacking each other in the 1780s.[65] Most striking of all, a recent analysis of the *cahiers* of the nobility in 1789 has shown them to be deeply divided ideologically, with large numbers of them won over to the political liberalism hitherto assumed to have been the monopoly of the bourgeoisie.[66]

How such findings affect the interpretation of the origins of the Revolution is not easy to assess, but attempts have been made to incorporate them into new explanations. The most ambitious to appear in the English-speaking world was that of Colin Lucas, published in 1973.[67] All recent work, Lucas argued, suggested that, 'The middle class of the later Ancien Régime displayed no significant difference in accepted values and above all no consciousness of belonging to a class whose economic and social characteristics were antithetical to the nobility.'[68] Bourgeoisie and nobles were all part of a single propertied élite. But this did not mean that there were no tensions within the élite, no 'stress zones' where tensions were explosive. Lucas took up Cobban's suggestion (largely ignored since it had been made) that the venal office-holders might occupy one such stress zone. Bypassed by the economic prosperity of the century, they found themselves unable to realize their social ambitions, and elbowed off the ladder of social ascent by richer mercantile newcomers. Their circumstances were strikingly parallel to those of the petty nobility, equally prevented by poverty from aspiring to what they considered their rightful share of public office and perquisites. These two groups together provided the driving force of hostility to the government in 1787–8. Why then did they not unite? Lucas found the answer in the declaration by the parlement of Paris in September 1788 that the Estates-General should meet according to the forms of 1614. For Lefebvre, too, this had been the crucial turning point, the moment when the bourgeoisie had seen that its interests were profoundly at odds with those of the nobility, the trigger which detonated the bourgeois campaign to capture control of the Estates.[69] This interpretation had been criticized in 1965 by Elizabeth A. Eisenstein on the grounds that the mysterious 'Committee of Thirty' which largely orchestrated the outburst of denunciatory pamphleteering which followed the parlement's ruling, had many noble members and therefore could hardly be considered the voice of the bourgeoisie alone.[70] Despite rejoinders from American defenders of Lefebvre that in objective terms the Committee of Thirty was still doing the bourgeoisie's work, Eisenstein's arguments lent strength to the tendency, culminating with Lucas, towards regarding nobles and bourgeoisie as part of the same social élite. But for Lucas the essential point about the parlement's ruling was not that it represented the unwillingness of a noble élite to share power with the bourgeoisie; it was rather that the distinction between nobles and non-nobles, obsolete now for several generations, was arbitrarily resurrected by the electoral requirements of the Estates-General, and at a moment when the lower echelons of the propertied élite were already suffering from economic and social frustration. Now they were faced with political frustration, too; so that 'The revolt of the Third Estate was a revolt against a loss of status by the central and lower sections of the élite with the approval of those elements of the trading groups which were on the threshold of the élite. It was this social group that became the "revolutionary bourgeoisie".'[71] Lucas, in other words, accepted Taylor's social and economic analysis but did not follow him into concluding that the Revolution was a purely political event without social causes. Nor did most others dare follow Taylor in this respect.[72] Yet even Lucas's 'stress zones' are looking increasingly unconvincing. By demonstrating that the price of most offices before the Revolution was rising, not declining, the present author believes he has finally reduced Cobban's 'declining bourgeoisie' to the status of a fruitful error.[73] That can only do severe damage to any interpretation, such as that of Lucas, which attempts to incorporate it. The effect is further to reinforce Taylor's position.

Even so, most historians continue to believe that the Revolution had significant social origins, although not necessarily the ones advanced by Lefebvre. Only isolated scholars now invoke the capitalist bourgeoisie as a revolutionary force. Yet many still share with Lefebvre the view that the essential achievement of 1789 was to remove the last obstacles to a recognition of what had happened in French society during the prosperity of the eighteenth century. The abolition of privilege and noble status was the destruction of the last hollow relics of a vanished social order: but it was not the replacement of feudalism by capitalism, or of the aristocracy by the bourgeoisie. Rather it meant that France was now to be ruled not by men of birth but by men of property, not by nobles but by 'notables'. Under this umbrella both former nobles and former bourgeois found a common shelter. The Revolution did not bring about this situation, it merely clarified it. This argument is not confined to the English-speaking world; it has also been persuasively advanced in France by Guy Chaussinand-Nogaret,[74] and is implicit in the work of Furet and Richet. A new international consensus appears to be emerging.

The oldest theory of the origins of the French Revolution is that it was some sort of intellectual conspiracy; the result, in some sense, of the Enlightenment. Bewildered contemporaries, alarmed at the unprecedented course of events, and unable to conceive of complex explanations, found comfort in the idea that the Revolution resulted from a philosophic or masonic plot. This became all the easier to believe when the revolutionaries quarrelled with the church, as the philosophers of the Enlightenment had done before them.[75] Nor have the sophisticated social and economic analyses of more recent historians entirely superseded such an interpretation. Occasional writers, mostly, it is true, not from among the ranks of professional historians, still see the Enlightenment as the Revolution's prime cause.[76] Even those who would deny it any such primacy feel obliged to admit that the climate of ideas must have played some part in the Revolution's origins. But there is little agreement on what precisely this role was.

The starting point of all modern discussion on this question is a monumental survey published by Daniel Mornet in 1933.[77] Though a professor of literature, Mornet tried resolutely to avoid the facile attribution of events to the influence of a few famous writers on a vague but unspecified audience. He believed that only a statistical approach to the diffusion of books and the ideas which they contained could yield reliable conclusions, and his research was based on such methods. His pioneering work has inspired most of the worthwhile research done in this field since then.[78] His conclusions were moderate. According to Mornet, the climate of ideas did not bring about the Revolution in any direct sense: that was the work of political factors. But the political events were influenced by a climate of opinion deeply dissatisfied with most aspects of the old order, and eager to demand wholesale reform as soon as the opportunity presented itself. This climate of opinion had evolved slowly, originating in attacks on organized religion during the first half of the century. In the years between 1748 and 1770, the society and the state which upheld the established church also came under comprehensive criticism, and by the latter date criticism had triumphed in polite society. Nothing was any longer sacred, nothing beyond discussion or criticism, and for the last twenty years of its existence the old order, while making little effort to respond to criticism, did little, either, to suppress it. As a result, disillusion with existing institutions and the way they worked spread down through society from the narrow, highly educated circles where it had originated, and out from Paris into the provinces, so that by the late 1780s much of France, at all social levels, was prepared for far-reaching changes and indeed eager to bring them about. There was, then, no philosophic plot; the great writers of the period were popular, and articulated discontents superbly, but they planned no revolution, and they were certainly no more influential than hosts of less memorable but just as prolific scribblers—among whom, incidentally, much the boldest and outrageous opinions were to be found. Nor was there a masonic plot: Freemasonry was a characteristic product of Enlightenment society, but most French masons, as masons, concerned themselves almost exclusively with masonic affairs. Those who played a more public role were scattered and unusual, and their efforts were quite uncoordinated. On closing Mornet the reader is left with the impression that events in France down to 1789 would have been very different if there had been no Enlightenment, for obviously the intellectual atmosphere would have been different, too. But that said, there was nothing uniquely dangerous or malignant about the thought of the eighteenth century, and it posed no serious threat to the old order until that order had begun to collapse for other reasons.

No new synthesis of this question has been attempted since Mornet, but most historians writing generally on the Revolution and its origins have felt obliged to take a view about it. Lefebvre's contention was that the Enlightenment was the ideology of the bourgeoisie; it spread as the bourgeoisie rose over the eighteenth century, and it triumphed with the bourgeoisie in the Revolution.[79] The enlightened emphasis on utility, rationality, individualism, and merit seemed to him to be the obvious product of the bourgeois mentality. The role of the *philosophes* had merely been to clarify these ideals and translate them into elevated language so that the bourgeoisie, in pursuing its own ends, could claim to be furthering the good of humanity as a whole. The Enlightenment did not cause the Revolution, but it was the programme of the social groups that did. This interpretation, however, at once tells us everything and nothing by implying that there is no real problem. Predictably, it failed to satisfy

Cobban, who saw the influence of the Enlightenment on the Revolution as far too sporadic and often too contradictory to represent a coherent programme, and he argued that in many ways the Revolution represented a reaction against what eighteenth-century thinkers had stood for.[80] Its political influence was far more evident in an 'age of reform' beginning around 1770, than it was in the Revolution;[81] and in *The Social Interpretation of the French Revolution*, Cobban ignored intellectual matters almost entirely. The revolutionaries, it seemed, acted almost exclusively from material motives. As if to emphasize this view with a case in point, in 1965 a pupil of Cobban's, Joan McDonald, produced a study of Rousseau's *Social Contract* which argued that it 'did not play an important part in shaping the views of those who participated in the events of 1789. It was only after 1789 that interest in the *Social Contract* began to develop.'[82]

Many English-speaking historians accepted such demonstrations with relief. They removed the need to set sail on the treacherous waters of the history of ideas, and left the historian free to give all his attention to more concrete phenomena, such as social and economic conditions.[83] Those who, like Norman Hampson, continued to emphasize the importance of autonomous intellectual convictions not directly related to the social circumstances of those holding them, remained isolated.[84] And although McDonald's arguments about Rousseau were shown in 1969 to be based upon insufficient evidence,[85] most recent research in English has tended to stress how limited was the impact of the latest ideas on the bourgeoisie.[86]

The final proof of the reforming desires of educated Frenchmen on the eve of the Revolution has usually been the evidence of the *cahiers* of 1789; Mornet concluded his great work with a chapter on them.[87] Many historians have been happy to echo the nineteenth-century judgement of de Tocqueville on the *cahiers*: 'When I come to assemble together all these particular wishes, I perceive with a sort of terror that what is claimed is the simultaneous and systematic abolition of all the laws and of all the customs obtaining throughout the kingdom.'[88] This was the harvest of speculative philosophy. Mornet was more cautious; most of the calls for reform, he believed, were of strictly practical inspiration, with no more than an overlay of philosophy derived from the isolated men of education who wrote out the *cahiers*. With George V. Taylor, however, caution had now turned to downright scepticism. The Revolution's initial manifesto, the Declaration of the Rights of Man and the Citizen, might have been cobbled together from a number of notions made familiar by the Enlightenment, but in no sense was the document as a whole the Enlightenment's product. It was, argued Taylor, the product of the political crisis in France. Nor was it in any sense a reflection of the general tenor of the *cahiers*. In his view, the most salient feature to emerge from a systematic survey of the *cahiers* of the Third Estate is their deep conservatism, and 'The only possible conclusion is that the revolutionary programme and its ideology were produced and perfected after the voters had deliberated in the Spring and that the great majority of them neither foresaw nor intended what was about to be done.'[89] A further obvious conclusion from Taylor's argument is that the *cahiers* contained even less of the Enlightenment than Mornet was prepared to concede, and that therefore its role in the origins of the Revolution was minimal.

Taylor's evidence, however, was drawn exclusively from the *cahiers* of the Third Estate. And no sooner had he argued that they showed little evidence of Enlightenment influence, than in France Guy Chaussinand-Nogaret concluded that the *cahiers* of the nobility were steeped in it.[90] Accepting Lefebvre's contention that the Enlightenment was fundamentally a bourgeois ideology, he argued that in the course of the eighteenth century large segments of the nobility were nevertheless won over by it. By the 1780s, they had abandoned traditional defences of their position and saw themselves as the natural leaders of a regenerated nation in which the way would be open to all on their merits. An analysis of their *cahiers* showed them to be deeply hostile to the old regime and anxious to create a state that was liberal, representative, and that protected and promoted the enterprising individual. Their reputed exclusivism in 1789 was a myth; in fact, what most nobles wanted was something strikingly similar to the aspirations of the bourgeoisie. This fits in well enough with Chaussinand-Nogaret's view that nobility and upper bourgeoisie had by this time in any case merged into one greater-propertied élite of 'notables'; but put beside Taylor's analysis (which Chaussinand-Nogaret did not appear to have read), it provokes interesting reflections. If the bourgeois programme took shape only after the elections, the nobility had elaborated it first; they must, therefore, have been more readily inclined towards radicalism than were the supposedly 'revolutionary' bourgeoisie. Nor did they learn their radicalism, as the bourgeoisie seemingly did, in the crucible of everyday politics between March and August 1789. They appear, in fact, to have got it from the Enlightenment, which was perhaps an influential movement after all. And this would certainly be the contention of Denis Richet, who argues that over the eighteenth century men of property, noble and bourgeois alike, found in the political theory of the Enlightenment powerful arguments to direct against a government that seemed both intent on overwhelming them with taxes, and too strong to be restrained from doing so by existing machinery.[91] Sales figures for the various editions of Diderot's *Encyclopédie* show that it sold well among both military nobles and professional bourgeois, precisely the most active and radical groups of 1789.[92] The movement of that year, in fact, is beginning to look like the work of a socially transcendent intelligentsia, homogenized by their education.[93] Accordingly, the ideas they had are beginning to be studied once more for their intrinsic importance rather than as mere rationalizations of social interests.[94]

To speak of a consensus on this issue would, certainly, still be premature. Nevertheless, only one aspect of the problem studied by Mornet still remains where he left it, with no serious historian inclined to dispute his conclusions. This is the issue of Freemasonry. Much important new work on eighteenth-century Freemasonry has been done in recent years,[95] which has emphasized its importance in the diffusion of certain 'enlightened' attitudes and in providing new, non-religious forms of 'sociability'. Isolated instances of political activity continue to be found, too, but nobody has resurrected the idea that Freemasonry played a conspicuous role in originating the Revolution. The 'pre-romantic' mysticism of which Freemasonry was one obvious symptom, however, has been shown to have a political importance all the same. In a study of the unpromising pseudo-scientific phenomenon of Mesmerism, a craze which swept fashionable circles in the 1780s, the American scholar Robert Darnton has shown that it was espoused by many ambitious intellectuals anxious to use it in order to make their names. When the Academy of Sciences, with the full approval of the government, pronounced the whole thing a hoax, its protagonists were outraged, and began to turn their thoughts towards criticism of the established authorities who had so blighted their hopes.[96] Subsequently, Darnton expanded his argument in a brilliant article.[97] He followed Mornet in arguing that by the last two decades of the old regime the 'heroic' Enlightenment had triumphed in France. Its doctrines were widely accepted in polite society, and its leaders had conquered the established institutions of knowledge and the intellect, such as the academies and the salons. But those who succeeded the Voltaires, the d'Alemberts, and the Diderots at the head of the movement when these giants died, and who inherited their social acclaim, had little new to say. And this provoked deep resentment among the 'literary rabble' of Paris who thought they did have something worth saying, whether about Mesmerism, institutional reform, or anything else. These swarming hacks hoped, like the great heroes of the Enlightenment before them, to write their way to fame and fortune. They found fame and fortune already monopolized by second-rate socialites who did not even put pen to paper most of the time, and yet who had the power and prestige to censor and condemn their works out of hand. Not unnaturally the *libellistes* of 'Grub Street' felt deeply hostile to an order which did such things to them, and although their resentment lacked a coherent programme, it found one when the Revolution replaced the old order and brought them, in the persons of men like Brissot, Marat, and Billaud-Varenne, to power. In short, the cut-throat competition within the literary world created an army of potential revolutionaries. The heroic Enlightenment had challenged a whole range of accepted values in its day, and played an important role in sapping the confidence in the established order of educated circles. But by the 1780s it no longer threatened anything. The danger now came from those who, regarding themselves as the continuators of the Enlightenment, were in fact being driven increasingly to reject what it had become.

Throughout the nineteenth and early twentieth centuries, historians were deeply divided about the economic origins of the Revolution. Few doubted that economic conditions were of fundamental importance in moulding the character of the events which reached their climax in 1789; but some claimed that these conditions were poor and miserable and getting worse, while others saw the pre-revolutionary years as a time of unprecedented prosperity. These disagreements, at least, have been resolved for the late-twentieth-century student by the work of Ernest Labrousse.[98] Labrousse's enormous statistical labours showed that between the 1730s and around 1770 the French economy made rapid and sustained advances, with good harvests, rising population, rising prices, and expanding overseas trade. But between 1770 and 1778, these prosperous times came to an end under the impact of a series of bad harvests, and there was no recovery over the next decade. On the contrary, difficulties continued, with wild, short-term fluctuations, and this period of uncertainty culminated with the truly catastrophic harvest of 1788. The most spectacular result was a swift and steady rise in the price of grain over subsequent months, a rise which reached its highest point on 14 July 1789. Wages, which already throughout the century had risen far more slowly than prices, did not follow this sudden jump in the cost of living; accordingly, there was a dramatic fall in the amount of money which wage earners were able to spend on items other than basic foodstuffs. Demand for manufactures fell, therefore, and the crisis spread to industry. Industrial production may have fallen by as much as a half between 1787 and 1789, and employers economized on labour costs by cutting wages or dismissing their workers. The political crisis of 1789, therefore, came at a time of high prices, falling wages, and mass unemployment; and this did much to explain the scale and intensity of popular violence, both in town and countryside, in the spring and summer of that fateful year. The Revolution broke out at a moment of severe economic crisis, and the popular intervention was the product of misery rather than prosperity. But it was not a long-endured misery, finally become intolerable; it was made more acute precisely by the sudden contrast with relatively recent prosperity.

No historian has seriously disputed Labrousse's analysis. It remains as definitive as Lefebvre predicted it would in 1937.[99] Whether, however, the economic crisis and the downfall of the old order were causally linked, or whether they merely coincided, is a matter on which disagreement remains possible. 'Whoever says economic crisis', declared Labrousse, 'at the same time says budgetary crisis, whether latent or overt. In an economic crisis, tax revenues fall off, receipts are reduced, public credit is weakened. By contrast, expenses rise with the cost of public assistance. So much so that in times of financial crisis governments experience a sort of instability, a particular vulnerability.'[100] More recently, the links between difficulties in the economy and the financial crises of the monarchy have been explored by J.F. Bosher,[101] and on this issue the last word has surely not been said. But there is also a broader perspective to this question. The crisis described by Labrousse illustrates some of the classic weaknesses of an economy of the type we have come to call under-developed. In such an economy, the agricultural sector is of overwhelming importance, and difficulties there have profound ramifications in all other sectors. Equally, agricultural inefficiencies prevent improvement on other fronts. If the economic crisis was of fundamental importance to what happened in 1789, it follows that France's political revolution largely resulted from the absence—in striking contrast with England—of an agricultural revolution.

Ever since the hostile observations of the English agronome Arthur Young on French agriculture in the late 1780s, the majority of historians have held that French agriculture was indeed backward. There was certainly much theorizing on agricultural topics between the 1750s and the 1770s, but it had few practical results. Nevertheless, in 1961, J.C. Toutain argued that the eighteenth century saw a rise in agricultural productivity so marked that it could have resulted only from revolutionary technical improvements. The yield would have been even greater had it not been for the constraints, such as internal customs barriers, which the Revolution swept away.[102] The unspoken implication was that the Revolution was an essential and necessary prerequisite of further agricultural growth; that it occurred because there *had* been an agricultural revolution rather than because there had not. Toutain's argument was based upon a formidable statistical apparatus, but in 1970 Michel Morineau produced an even more monumental reply, which argued that the traditional view was right after all.[103] Mistrusting Toutain's sources, he argued that the levels of productivity revealed in an official national survey of 1840 were no higher than those which could be demonstrated for a century beforehand. By 1740, northern and western areas of the country were relatively advanced in their agricultural techniques, while the south and centre lagged far behind; but no change of any significance occurred in either area over the century studied, and so there had been no agricultural revolution. Morineau even anticipated the first line of defence of those who believed there had been one. They would ask, he foresaw, how the population of France could have risen so dramatically over the century—from around 19 millions in 1700 to around 25 millions in the 1790s—if agriculture had not revolutionized its techniques to feed so many more mouths. His reply was that the population level of the beginning of the century was well below the capacity of an unrevolutionized French agriculture to sustain, and that the rise over the century was far from dramatic in comparison with that of other countries. Not only was there no agricultural revolution; there was no 'demographic revolution' either. The growth in population over the eighteenth century was a mere restoration of the losses of the seventeenth, to levels which might well have already been attained in the sixteenth.[104]

Since Morineau's intervention the figures for France's population growth over the century have been authoritatively revised. The population in 1700 now appears to have been around 21 millions and that in 1789 more like 28.[105] But the order of rise remains the same, and Morineau's conclusions about how it was sustained have serious implications for our view about the economic significance of the Revolution. Lefebvre's generation, and Labrousse's, too, never doubted that the Revolution was an event of major economic significance. It was the point at which France, economically as well as socially, changed course towards the modern world. In this sense it was unique and unprecedented. Morineau, however, emphasizes the continuity of French economic history; he implies that nothing really new was happening in the eighteenth century, and that only much later did decisive structural modernization begin to occur. The existence of Labrousse's crisis is not in doubt, but it was not so much the final breakdown of an economic structure riddled with fatal contradictions: it was simply the result of a series of meteorological accidents, the sort of crisis that had often occurred before. It passed, and the basic economic structure of the country remained much the same as before. More recently, the implications of these ideas have been clearly spelled out by Roger Price.[106] What came to an end in 1789, he argues, was the *political* old regime. The *economic* old regime went on until the 1840s, when the advent of the railways at last began to break down the regionalism and lack of available markets that had always held back French agriculture in the past.

New perspectives of this sort only become possible by taking a long-term view of French history. Too often economic and social trends apparent in the years immediately before 1789 have been accepted by historians of the Revolution as new and peculiarly significant, when a longer perspective reveals them to be phenomena that had occurred before, or were indeed built into the very structure of the old order. Thus it has now been argued that the so-called 'feudal' or 'seignorial reaction', by which lords were reputed to have increased the burden of seignorial dues payable by their peasants in the pre-revolutionary decades, may well be an illusion.[107] None of this is to deprecate the sense of hardship or injustice that the peasantry may have felt towards the old order on the Revolution's eve, or to suggest that they had nothing serious to feel aggrieved about. It merely implies that there was nothing new in kind about the difficulties against which, in the summer of 1789, popular anger exploded. But undoubtedly such conclusions do rob the Revolution of much of the general significance which early twentieth-century historians attributed to it. No longer is it the cataclysm in which one economic order came to an end and a new one dawned. It is merely the last great crisis of a type to which the old economic order was peculiarly prone, but one which that order survived, to die finally, not with a bang but a whimper, half a century later.

Against those who take this view, of course, there is an obvious criticism. If the upheavals of 1789 were merely the result of the type of economic crisis that had been seen in 1693/4, in 1708/9, in 1740/1, or in 1770/1, why were subsequent events so very different this time, and why did previous crises not result in events comparable to those of the 1790s? Such a challenge is understandable, but it springs from the debatable assumption, which ran deep in the writings of early twentieth-century historians from Jaurès to Soboul, that a political event on the scale of the French Revolution must have had profound social and economic causes. And this assumption went even further: the social and economic causes were the important ones, the ones to do research on. The political origins of the Revolution, in contrast, were of marginal interest or importance; they were the foam on the great waves of history, little more than the 'histoire événementielle' dismissed so scornfully by the father of modern French economic history, François Simiand.[108]

In some ways the rejection of political history by early twentieth-century students of the Revolution was entirely healthy. The exhaustive and minute concentration by nineteenth-century scholars on the narrative of political events often contributed little towards understanding the deeper forces which moulded and sometimes determined their course. The social and economic research into the Revolution and its origins which resulted from this great reorientation of interest has immeasurably enriched our understanding of these things—even if it has not always clarified matters in the way that its more optimistic advocates perhaps hoped. But if the assumption, that only the social and economic structures of pre-revolutionary France are worth the historian's attention, is not accepted, then the neglect of political history by the early-twentieth-century masters (and their pupils) appears little less than lamentable. If, as George V. Taylor argues, the origins of the Revolution were after all primarily political, the politics of the old order cease to be the trivial and superficial squabbles of a worthless ruling class inexorably doomed by the relentless march of History. They become, after all, fundamental to an understanding of the Revolution's origins.

Interest in the politics of pre-revolutionary France has been slow to reawaken from the oblivion into which it fell in the early twentieth century. In the absence of new research, or of any awareness of the need to rethink established notions, historians were content to repeat a view of the Revolution's political origins first formulated far back in the nineteenth century in monarchist circles. Understandably, post-revolutionary monarchists were mainly concerned to see whether the upheavals of the 1790s might somehow have been prevented.[109] The best prospects for this, they believed, had lain in a reforming monarchy, and they thought they saw plenty of evidence between the 1760s and the 1780s that the monarchy was indeed eager to introduce reforms. But these reforming instincts, which might have brought about most of the salutary changes later wrought by the Revolution, without its violent excesses, were thwarted by the same enemy that the Revolution itself had more success in overcoming—the nobility. The old regime was destroyed by reactionary aristocratic opposition to a benevolent reforming monarchy. And the spearhead of this aristocratic obstructionism was the *noblesse de robe* of the sovereign courts of justice, the parlements.

Throughout the first sixty years of this century this interpretation went largely unchallenged. The work of Marcel Marion, the last great institutional historian in the nineteenth-century tradition to write on the pre-revolutionary years, is drenched in hatred for the parlements and uncritical applause for any change proposed by the crown.[110] The very popular writings of right-wing admirers of strong government, like Pierre Gaxotte or François Piétri, also propound such a view,[111] and Lefebvre, from a very different political stand point, found nothing to disagree with in it.[112] After the war it was given renewed popularity in France in works of popularization,[113] in new celebrations of the virtues of authoritarian ministries,[114] and even in monuments of profound scholarship on monarchical institutions.[115] English-speaking readers were introduced to it through the writings of Alfred Cobban, who accepted it completely.[116] These writers, and others like them, were not unanimous about when the old order reached the point of no return. Some believed it came with the fall of the 'enlightened' minister Turgot in 1776;[117] while others, perhaps most, saw the fatal moment as the dismissal of Chancellor Maupeou, and the restoration of the parlements he had muzzled and remodelled, in 1774. 'Maupeou' wrote Cobban, 'had. . .dramatically and successfully restored royal authority after a generation of weakness. Given a few more years for the country to appreciate the benefits of the new system, and there could hardly have been any question of a restoration of the parlements. Freed from their incubus. . .reforms. . .could be brought about and applied. The tragedy was that the few years that were needed were not given.'[118]

Nothing in the first modern piece of research on the parlements appeared to contradict this view. This was a massive study by Jean Egret of the political role of the parlement of Grenoble in the latter half of the eighteenth century; it was published in 1942.[119] But twenty-eight years later, towards the end of a lifetime devoted to their study, Egret produced a general survey of the parlements' activities under Louis XV which struck a note of sympathy not heard for generations.[120] So far from being the selfish and reactionary oligarchy conventionally portrayed, the magistrates of the sovereign courts played an important part in the political education of the pre-revolutionary nation, and they articulated many genuine grievances. Maupeou had indeed silenced them, but his reforms were limited in scope and brought the government into disrepute; in any case, after their restoration the parlements never caused the government the same trouble as they had before 1771. By this time doubts about the accepted view had also begun to be heard from across the Channel. In 1968, for instance, in a general history of the parlement of Paris, J.H. Shennan stressed the role of the parlements as defenders of the law and the rights of the subject against the authoritarian tendencies of the crown.[121] Then in 1970, unaware that Egret was about to produce similar views, I argued that Maupeou was not a serious reformer, but that his importance in French history was to demonstrate the weakness rather than the strength of the parlements. The parlements did not prevent the government from making reforms; the government had neither the will to make them nor the perception to see that they were necessary. The old order was brought down by a collapse of confidence among the financial community in its ability to manage its affairs, and not by the strength of noble opposition.[122] These ideas, however, appear to have made no impact in France, while outside they have either been rejected[123] or received with little more than polite but cautious interest.

The most successful challenge to traditional historical judgements on the political origins of the Revolution since 1939 has been the recovery of the reputation of Calonne. Lefebvre, in 1939, clearly thought there was more to be said for his reform plan of 1787 than previous generations had allowed,[124] and that year also saw an American attempt to rehabilitate the economic policies that Calonne had pursued since coming to power in 1783.[125] In 1946, Albert Goodwin produced a detailed analysis of his reform plan and the way it was shipwrecked in the Assembly of Notables.[126] With his fall, Goodwin concluded, 'there had now been removed from French politics the one man who might conceivably have healed the financial disorders which were shortly to ensure the collapse of the *ancien régime*.'[127] The plan was also favourably treated by Jean Egret in 1962, in perhaps the most important work on pre-revolutionary politics to appear this century,[128] but Egret did not see Calonne's fall as reform's last chance. His successor, Loménie de Brienne, for long depicted as an ambitious, unprincipled cleric who flitted ineffectually through the twilight of the old order, was now seen to be a thoughtful and serious reformer who failed more through bad luck and bad timing than because of the strength of opposition.

Much of Calonne's recently recovered prestige rests upon the willingness of historians to take the copious self-justifications which he wrote after his fall at their face value. To do this, they are obliged to reject the no-less-plausible apologias of Necker, since most of these writings were the product of controversy between the two men about their respective conduct of financial affairs. Low as Calonne's reputation has sometimes sunk, it has never sunk as far as that of Necker. If several figures—whether Maupeou, Turgot, Calonne, even Brienne—have been pushed forward for the rather dubious accolade of 'the one man who might have saved the old regime', Necker has always been almost the only contender for the title of the one man who destroyed it. Necker was the minister who, in financing France's part in the American War of Independence entirely through loans, left the monarchy with a crippling burden of debt. Worse, he compounded this crime by claiming, in his *Compte rendu* of 1781, that the royal finances were in modest surplus. Back in office in 1788, it was he who opened the floodgates to Revolution by conceding double representation for the Third Estate in the Estates-General; and in June 1789, it was he who torpedoed the belated royal reform proposals by conspicuously absenting himself when they were presented. The fact that he was a Swiss banker—a foreigner and a profiteer—did nothing to make chauvinistic French historians any more sympathetic towards him,[129] and even his biographers felt constrained to admit that his influence on the course of pre-revolutionary events was unfortunate.[130]

Yet no historical judgement is beyond dispute, and in recent years Necker's reputation has begun to recover quite dramatically. It has been argued that during his first ministry he introduced radical reforms in financial administration, which, had they not been abandoned by Calonne, might have given the system the strength to ride out the storms of the later 1780s.[131] His more general ideas on politics and the reform of the state have begun to be recognized as at least as worthy of attention as the much more celebrated views of Turgot.[132] It has now been plausibly argued, too, that his *Compte rendu* of 1781 was not a false picture of the financial situation, and that the loans he raised during the American

War were neither as large nor as catastrophic in their legacy as has usually been claimed.[133] Since much of these writings reflect their authors' readiness to believe Necker's version of affairs rather than Calonne's, they appear to signify that the latter's star is about to wane once more. Necker, meanwhile, the foreigner who had so little sympathy with the ways of the old order, seems ready to enter for the first time that select company in which he, at least, would have been quite unsurprised to find himself, of potential saviours of the old regime.

No survey of this length could possibly mention, much less analyse, all the changes in historical judgement about the origins of the French Revolution that have occurred since the 1930s. It has, for instance, ignored controversies over the idea, first propounded in the 1950s, that the Revolution was the result of trends found not only in France but throughout the whole west European or 'Atlantic' world.[134] No mention has been made, either, of new views about the intelligence and intellectual attainments of one of the central figures of the period, Louis XVI himself.[135] But enough should have been said to demonstrate that all the apparent certainties of forty years ago have dissolved. Disagreement, controversy, and iconoclasm are to be found in all corners of the field.

If these disagreements were about the same material, there would be grounds for despair that we should ever again feel sure why the French Revolution occurred. But happily this is not the case. The fruitful thing about historical controversy is that it suggests new questions and stimulates fresh research in order to answer them. Thus the present-day historian has incomparably more material upon which to formulate his judgements about the Revolution's origins than Lefebvre or his contemporaries had access to, material that largely owes its existence to the stimulus of the debates that have been chronicled here. This does not necessarily make the task of synthesis any easier; quite the reverse. But it does suggest that a new synthesis, however difficult, however imperfect, however tentative, is urgently required.

PART II

A. The Breakdown of the Old Regime

1. The Financial Crisis

The revolution that was to sweep away the political institutions of old France, and shake her society to its foundations, did not begin on 14 July 1789. By that time the old order was already in ruins, beyond reconstruction. This was the result of a chain of events that can be traced as far back as 20 August 1786. For it was on that day that Calonne, comptroller-general of the royal finances, first came to Louis XVI and informed him that the state was on the brink of financial collapse.

We have no absolutely reliable or completely unambiguous figures to illustrate the financial condition of France in 1786. Nor did contemporaries have such figures.[1] Even Calonne, with all the accounts of the royal treasury at his disposal, claimed that it had taken him two years to arrive at his own assessment of the problem. But the seriousness of the situation was beyond dispute.[2] According to Calonne, the total revenue for 1786 would amount to 475 million *livres*, but expenditure would probably total 587 millions—a deficit of 112 millions, or almost a quarter of the annual revenue. When Louis XVI had come to the throne in 1774, Calonne claimed, the deficit had been 40 millions, and it had even fallen over the next two years. But since 1777 it had risen steadily, and there was every prospect, over the next few years, of its rising further if drastic action were not soon taken. The basic reason for this deterioration was that since 1777 there had been an enormous rise in state borrowing and consequently in the annual interest and repayments that the treasury was obliged to disburse. Since 1776, Calonne claimed, 1,250 millions had been borrowed. Until 1794, 50 millions per year of short-term loans would fall due for repayment, and meanwhile, the cost of servicing the total debt ate up nearly half of the annual revenue. Worse still, no less than 280 millions of the next year's revenues had already been anticipated in order to raise money for earlier expenditure.[3]

Financial difficulties were nothing new under the French monarchy. Indeed, throughout the seventeenth and eighteenth centuries they were the normal state of affairs; it was the rare moments of financial health that were extraordinary.[4] Nor was the cause of these difficulties any mystery. Successive kings had always spent too much on war. The wars of Louis XIV had imposed a crippling legacy of debt on the royal finances, and although this burden was much alleviated by the great financial crash of 1720/1, which enabled the government to write off huge sums, four major European and overseas wars since that time had brought matters once more to crisis proportions. They were already serious by 1763, at the end of seven years of costly and unsuccessful conflict on a worldwide scale; the debt had almost doubled in a decade,[5] and for the next fifteen years successive comptrollers-general of the finances warned unceasingly against the dangers of further wars. French participation in the American War of Independence between 1778 and 1783 was glorious and successful, but it confirmed these ministers' worst fears. By 1783, the financial situation was as bad as it had been in 1715, and over the next three years it continued to deteriorate to the point which Calonne announced to the king in August 1786.

A number of obvious expedients are open to governments in financial difficulties. Unfortunately, most of these expedients were not open to Calonne—or, if they were, there were reasons why they could not prove as effective as they should.

One natural step, for example, was to effect economies. There was undoubtedly scope for this, and the plan which Calonne put forward to the Assembly of Notables the next year was to include a number of money-saving proposals. Nevertheless, none of the major items of public expenditure could be substantially reduced. The greatest of them—debt service—could only be diminished by reducing the capital of the debt. Calonne had indeed begun this process by establishing a sinking fund (*caisse d'amortissement*) in 1784, in the hope of paying off 3 millions a year. But in any case many short-term debts were due for repayment in the decade beginning in 1787, and it was precisely this impending outlay that made the position in 1786 so serious. The second-greatest item, the armed forces, could only be markedly reduced at the cost of jeopardizing France's international position at a moment when the internal instability of the Dutch Republic and uncertainties in Eastern Europe following the death of Frederick the Great made the international situation ominous. Economies, therefore, must largely be a matter of trimming expenditure over a whole range of minor items such as pensions, the royal household, public works, and welfare services which together accounted for only about one-seventh of annual outlay.[6] In 1776, the severe Turgot had only been able to envisage savings of 34 millions, mainly from such sources,[7] which was nowhere near enough to meet a deficit on the scale of 1786. Clearly economies could only be effective in conjunction with more comprehensive measures.

A second possibility would be to increase taxes. Yet France already thought of itself as one of the most highly taxed nations in Europe. It is true that the average Dutchman or Englishman paid more per head in taxes than his French counterpart; but in France there were immense regional diversities, so that taxpayers in the Paris region paid more per head than anybody else in Europe.[8] And when we consider that the populations of Great Britain and the Dutch Republic were in any case far wealthier as a whole than that of France, the French burden appears all the greater. What is more, it seemed to have increased inordinately within living memory. Whether the real weight of taxation in 1786 was much greater than in 1715, given the economic growth of the intervening years, is not at all certain.[9] In cash terms, however, an increase seemed obvious. In 1749, a new tax on landed property of 5 per cent had been introduced—the *vingtième*. It proved to be permanent. In 1756, it had been doubled for a limited period, but in practice the government never felt able to do without the extra revenue, so this second *vingtième* became in effect as permanent as the first. Between 1760 and 1763, the most costly period of the Seven Years War, a third *vingtième* was levied; and in 1782, it was reintroduced with the assurance that it would end three years after peace was concluded. That moment came at the end of 1786, and this imminent fall in revenue was another of the factors which led Calonne to confront the crisis when he did. There had also been substantial rises in indirect taxation, especially under the ruthless Terray, who between 1770 and 1774 added 60 millions to the revenues. Terray's memory was execrated right down to the Revolution, and by none more than landowners and nobles, who had borne the brunt of innovations such as revisions of the *vingtième* assessments, new taxes on venal offices, and duties payable on assuming nobility. It is true that even in 1786 direct taxation was not a major burden on noble incomes, especially in comparison with non-noble ones.[10] But to nobles the rise since 1749 (before which they had paid only one minor permanent direct tax, the *capitation*) still seemed enormous, and as the wealthiest and most articulate section of society, they were best placed to protest against the increasing burden. The repeated lamentations of landed, noble bodies like the parlements made the matter a constant topic of public concern. This atmosphere did much to induce Necker to finance the American war by borrowing rather than increased taxation, but the very fact that this had been possible made new taxes introduced after three years of peace seem less acceptable than ever. So it was politically, psychologically, and perhaps even physically impossible to increase the over-all weight of taxation. All that could be done was to redistribute the burden so that it fell more equitably and was levied with more accuracy.

Another possibility was for the state simply to renounce its overwhelming burden of debt by declaring bankruptcy. Earlier governments had often adopted this expedient; but over the eighteenth century it had come to seem less and less respectable. The financial crash of 1720, in which thousands of government creditors were ruined, and a series of reductions in the *rentes* (government annuities) in the chaotic years following that crisis, had instilled French public opinion with a deep hostility to breaches of public faith, and from 1726 onwards governments had striven to keep public confidence by avoiding any suggestion that they might default on their debts.[11] But circumstances were not always entirely within the government's control, and in 1770, amid the most serious economic crisis for decades, Terray had felt compelled to suspend payment of short-term credits and reduce or defer payment of other government debts.[12] These operations coincided with Chancellor Maupeou's ruthless reorganization of the judiciary,[13] and there was a general outcry against an arbitrary government which thought nothing of renouncing its most binding obligations. The government's credit was shaken as investors concluded that their money was not safe in its keeping, and Terray had difficulty in covering the new loans that he floated. To subsequent ministers the lesson seemed clear; bankruptcy was not only dishonourable, it destroyed the state's credit and made further borrowing difficult. Terray's successor, Turgot, and every minister who followed him, Calonne included, set their faces firmly against even partial bankruptcy. The determination of successive revolutionary assemblies down to 1797 to honour the debts accumulated under the old order shows how deeply and generally public opinion shared the view that the public debt should be sacrosanct.

But Calonne could hardly go on borrowing. It is true that the plan of action which he laid before the king contained proposals for further loans in order to cover the repayments of the coming years, but there was no long-term future in fighting a debt problem by new borrowings. In any case it was uncertain whether such borrowing would even be possible. The French government already borrowed money on terms distinctly less favourable than either the British or the Dutch, because in France there was no publicly supported bank through which government credit could be cheaply channelled. Under the Regency, the Scottish adventurer John Law had tried to set up such a bank, but by linking it too closely with the fortunes of his ill-fated Louisiana company, he ensured that it collapsed in the crash of 1720. The legacy of this experience was a deep public hostility in France both to state banks and the paper money that they issued. When in 1776 the experiment was renewed, in a limited way, with the establishment of the *caisse d'escompte* (discount bank), the institution was rapidly taken over by speculators rather than by the investing public at large, and the help it gave to the government's credit was marginal.[14] In the absence of a bank, the government was compelled to rely on intermediaries for raising its loans, bodies like the municipality of Paris, the estates of provinces that retained them (like Languedoc and Brittany), or great corporations like the clergy, all of which could borrow money on better terms than the king. But when needs were extraordinary even these resources were not enough. Then, the government was compelled to float loans on its own behalf, but on terms so generous that even the most prudent investor found them hard to resist. This is what happened between 1777 and 1786.

Conditions of credit were already difficult throughout this period. The years around 1770 witnessed a European economic crisis whose ramifications shook all the great financial centres. Terray's partial bankruptcy, which damaged government credit particularly badly, was one result of the crisis. In a sense every major expedient adopted by the successive ministries of Louis XVI was designed to bolster or restore a shaken public confidence in order to keep the loans flowing. This consideration certainly played its part in the dismissal of Maupeou and Terray and the reversal of their policies in 1774. It was important, too, in the appointment of Turgot to succeed Terray, and when he fell two years later, government stock lost 8 or 9 per cent.[14] It was certainly the main reason why Necker, a well-connected banker, was put in charge of the royal finances in 1777. The appointment coincided with the approach of war against Great Britain, and the need which such a war would undoubtedly bring for massive new loans. Necker did not fail: between 1777 and 1781 he raised 520 million *livres.*[15] But most of this money was borrowed for relatively short periods, less than twenty years, and offered a return of anything up to 10 per cent per year. All this made Necker's loans extremely profitable ways of investing money. It also made them ruinously expensive to the state; but the potential economic gain of shattering the British Empire seemed ample justification for such a special war effort. And so all Necker's loans were subscribed within days. He was even able to persuade many of the normally cautious investors of Holland to forsake their long-standing preference for the British national debt and lend money to the French king.[16] His success meant that he was able to achieve what had hitherto been thought impossible—to finance a major war without any new taxation. And in 1781, lest this achievement strain public credulity, he issued the first public balance sheet of the French royal finances, the famous *Compte rendu au Roi*. His objective was to sustain confidence by demonstrating that the king's ordinary accounts were in modest surplus. He said nothing about *extra*ordinary accounts, from which the bulk of the war effort was financed; but nobody noticed this omission.[17] Necker preserved his reputation as a miracle worker. The blow to confidence was all the greater, therefore, when a few months later political reasons brought about his resignation. Only by continuing to borrow heavily, and by introducing at last extra taxation such as the third *vingtième*, were Necker's two immediate successors able to sustain credit; and even then by the time Calonne took over in 1783, as he eloquently put it, 'All the funds were empty, all public stocks were low, all circulation was interrupted; alarm was general and confidence destroyed.'[18] Calonne saved the situation by a massive programme of public expenditure, which was intended to demonstrate that the government was not worried. He spent lavishly on what he called 'useful splendour'—the court, royal palaces, and major military projects like a new naval harbour at Cherbourg. For a time this tactic succeeded, and between 1783 and 1787 Calonne was able to borrow no less than 421,788,660 *livres*—only about a hundred millions (or 20 per cent) less than Necker had.[19] It bought him the time he needed to make a thorough examination of the financial situation, but it made that situation far worse by the moment he decided what to do. By 1786, few people believed that borrowing on the scale of the previous ten years could go on. The parlement of Paris, for instance, had registered all Necker's loans without demur, but protested at those floated by Calonne in December 1784 and December 1785. And once through the parlement, these loans were still only subscribed very slowly, which showed that the financial world, too, now feared that the government was overextended. At the same time the long economic recession, which had begun around 1770, reached a new crisis in 1786, a depressed economy producing low tax yields, which in their turn put pressure on the government's main security for its credit, tax revenue.[20] So that even if Calonne had wished to continue borrowing in 1786, the prospects for success were poor—unless he could engineer some wholly novel boost to confidence.

An expedient that Calonne did not consider was a reform of the system by which the government financed its activities.[21] One reason why it took him so long to come to any conclusions about the true state of the finances was that the king had no central treasury where accounts were kept, revenues taken in, and payments made. Nor was there any real notion of an annual budget. Most of the state's finances were handled by independent financiers who had bought the right to handle government revenues, either through membership of the company of Farmers-General, who collected most of the indirect taxes, or through buying an office of accountant (variously called payers, receivers, or treasurers) to a government department. Once in office, all that these officials were obliged to do was to receive or pay out funds on the government's orders, and send in periodic accounts to the crown's courts of audit, the *chambres des comptes*. What they did with the money in their accounts otherwise was their own affair. And what they often did with it in practice was to lend it to the government in short-term credits—so that the king found himself borrowing and paying interest on his own money. The day-to-day payments of the government, in fact, depended on short-term credits of this sort, the *anticipations* which ate up so much of the expected revenue for 1787, advanced by men who were nominally state employees, but who in reality were private businessmen making a profit from manipulating public funds. Nor did such businessmen confine their activities to juggling with the state's money. They normally had extensive private financial dealings, too, and made no distinction between the two fields of activity. So that when, in times of economic stringency, their operations came under strain, so did the finances of the government. This is what happened in 1770 and again in 1786/7, when the government's difficulties were heralded by the bankruptcy of a number of its financiers. A state bank would, of course, have freed the government from its dependence on these profiteering agents, who constituted in effect a body of several hundred petty bankers. They knew this very well: they had been the mainstay of opposition to John Law's Royal Bank under the Regency; they were the main beneficiaries of its collapse, and the main source of arguments against repeating the experiment right down to the Revolution. Alternatively, the government could have saved itself substantial sums by abolishing the system whereby it depended for its operations on private profiteers, and consolidating the whole machinery of royal finances within one central public department handling all revenues and all payments, staffed by full-time, salaried public officials. In the wake of the difficulties of 1770, Terray began to move in this direction by abolishing a number of financial offices. Turgot abolished several more, and Necker began to cut them down wholesale. At the same time Necker consolidated many disparate government funds into a few large aggregate ones, and tried to impose an unprecedented degree of day-to-day supervision over the management of the state's finances from the centre. But these efforts came to an end when he fell in 1781. Calonne believed that Necker's systematic attack upon the role of private financiers in the government's affairs alienated them and therefore harmed the state's credit in circles on which it had traditionally relied. He restored much of the old system, and in 1786 was thinking more of strengthening than of destroying it, convinced that the cost of reforming it would far outweigh any immediate benefit in regular savings.

With so many courses of action either closed or considered impractical, what then did Calonne propose to do in order to resolve the crisis? Nothing could put it more clearly than his own words. 'I shall easily show,' he declared to the king, 'that it is impossible to tax further, ruinous to be always borrowing and not enough to confine ourselves to economical reforms and that, with matters as they are, ordinary ways being unable to lead us to our goal, the only effective remedy, the only course left to take, the only means of managing finally to put the finances truly in order, must consist in revivifying the entire State by recasting all that is vicious in its constitution.'[22] He was proposing something quite unprecedented in the history of the monarchy—a total and comprehensive reform of all its institutions, according to clear principles, in such a way that it should never fall into difficulties like those of the 1780s again.

In the document he presented to the king, *Summary of a Plan for the Improvement of the Finances*, Calonne never defined his guiding principle in a few words. But it emerged very clearly from the way he put the problem.

The disparity, the disaccord, the incoherence of the different parts of the body of the monarchy is the principle of the constitutional vices which enervate its strength and hamper all its organization;. . .one cannot destroy any one of them without attacking them all in the principle which has produced them and which perpetuates them;. . .it alone influences everything;. . .it harms everything, . . . it is opposed to all good; . . . a Kingdom made up of lands with estates, lands without [*pays d'élections*], lands with provincial assemblies [*administrations provinciales*], lands of mixed administration, a Kingdom whose provinces are foreign one to another, where multifarious internal barriers separate and divide the subjects of the same sovereign, where certain areas are totally freed from burdens of which others bear the full weight, where the richest class contributes least, where privileges destroy all balance, where it is impossible to have either a constant rule or a common will, is necessarily a very imperfect kingdom, brimming with abuses, and one that it is impossible to govern well; . . . in effect the result is that general administration is excessively complicated, public contributions unequally spread, trade hindered by countless restrictions, circulation obstructed in all its branches, agriculture crushed by overwhelming burdens, the state's finances impoverished by excessive costs of recovery, and by variations in their product. Finally, I shall prove that so many abuses, so visible to all eyes, and so justly censured, have only till now resisted a public opinion which condemns them, because nobody has attempted to extirpate their germ, and to dry up the source of all obstacles by establishing a more uniform order.[23]

In short, the French state lacked rational organization and uniform principles, and it was not enough to attempt to solve financial problems, as previous ministries had, by exclusively financial means. Calonne believed that it was now necessary to reform the economy, government, and to some degree French society itself. The details of his plans are best considered later. Louis XVI was staggered by their scale, and it took some months to convince him that they were really necessary. Under Louis XIV, to convince the king might have been enough to ensure that a plan was implemented. But the governmental processes of the monarchy had grown more complex since the Sun King's time.

2. The System of Government

The king of France was an absolute monarch. This meant that there was no institution in the state with the right to prevent him from doing whatever he chose to do, in contrast to a state like Great Britain, where royal power was circumscribed by Parliament. It is true that there were certain 'fundamental laws' which the king was expected to observe, such as those governing the succession. But there was no consensus for the most part over which laws were and which were not fundamental; and in any case a determined king could override even those which did command general recognition—as when Louis XIV placed his two bastard sons in the line of succession. As the mid-eighteenth century chancellor, Lamoignon, declared, the king of France was a sovereign to whom everything was not permitted, but everything was possible.[1]

In this sense the king bore the final responsibility for everything that the state did—or did not do. And if the old regime failed to reform itself, this was in large measure the fault of Louis XIV, Louis XV, and Louis XVI, for they alone had the power, the authority, and the right to authorize necessary reforms. Of these three, only Louis XIV was a monarch of superior abilities, who in his diligence, regular habits, fixity of purpose, and firmness under pressure set new standards of monarchical conduct. His policy was to improve and maintain his own authority in the state, but he never thought of using it to make sweeping reforms. Indeed, he was responsible for much subsequent governmental confusion in that he seldom abolished institutions which he deprived of power. He thought it enough merely to transfer their powers elsewhere, which worked so long as he was on the throne, but opened the way to endless conflicts of jurisdiction and paralysing rivalries under less assured rulers. Louis XV was an intelligent man, whose concern for his responsibilities has often been underestimated by historians.[2] But he was shy, idle, and easily bored compared with his predecessor, and he preferred to avoid problems, or seek stopgap solutions, rather than to lay comprehensive, long-term plans. Like Louis XIV he was more interested in protecting his authority than in using it imaginatively. Louis XVI was more ambitious than this. He wished to be popular, and he wished to use his power benevolently; but he had no clear idea of what this entailed, and he mistrusted his own judgement.[3] Great projects and bold initiatives could not be expected from such a ruler, and despite his determination in 1774 to break with the bad old ways of his predecessor, Louis XVI saw no necessity for radical departures from established procedures or priorities until Calonne persuaded him otherwise in the autumn of 1786.

Louis XIV had deliberately imprisoned himself at Versailles. He hated turbulent, rebellious Paris, and he wished to create an enormous stage for the monarchy where all the most powerful men in the kingdom could be concentrated under the king's eye. In this way he was able to tame and domesticate the higher nobility, hitherto the most formidable threats to his authority. But this policy had its price. The king largely isolated himself from firsthand knowledge of his kingdom, and only saw provincial France when he went campaigning. And neither of his successors broke away from this pattern. Louis XV spent less time at Versailles, but only because he was constantly touring around other palaces in the Paris region. He saw more of Paris but seldom went to the front. Louis XVI's only venture into provincial France (before the flight to Varennes in 1791) was a journey to Cherbourg to view the new naval harbour in 1786. The values and narrow outlook of the petty, vicious, overheated world of the court could hardly fail to influence the monarchs at its centre, or to warp their view of life outside. Louis XV and Louis XVI, unlike the architect of the system, had never known anything else, and never showed any sign of thinking that anything else might be more desirable.

The rulers of eighteenth-century France, then, were intelligent, but not men of ideas or initiative; in contrast to contemporaries like Joseph II, Gustavus III, Frederick II, or even (according to his lights) George III. The French kings depended far more than many others on the advice they received from their counsellors. Accordingly, the quality and effectiveness of royal policies depended in its turn on who these counsellors were. Every monarch recognized the obligation to seek advice before acting, for this was one of the most hallowed and fundamental of French constitutional principles. Louis XV was always recalling with approval the dying advice of Louis XIV to take counsel in all things.[4] The essence of Louis XIV's seizure of power in 1661 had been his decision to be advised only by men of his own choice, in preference to those who claimed an hereditary or ex-officio right to sit on the royal council. Louis XV and Louis XVI remained true to this principle, but neither they nor their great predecessor profited from their freedom of choice to consult new men or to seek advice outside established circles. Louis XIV inherited his most trusted advisers from Mazarin, kept most of them in office until they died, and then often turned to their sons. Louis XV was content to leave government in the hands of his tutor, Cardinal Fleury. When the latter died in 1743, Louis was in his early thirties, and he selected his advisers from then until his death from the only circles he knew—from courtiers (which Louis XIV had never done), from the administrative hierarchy, and from the courts of law. Louis XVI did the same—except in the case of Necker, the foreign Protestant banker. Necker served his purpose, in that he secured the loans that the government needed; but he had ideas and plans for recasting governmental institutions and practices beyond recognition,[5] and his unorthodoxy eventually cost him his place.[6] The upshot was to turn the king back towards the established fields of recruitment, and the ministry which confronted the financial crisis in 1786 was an entirely traditional blend of courtiers, career diplomats or administrators, and magistrates. They were united (in this if in little else) by the fact that they had all come to the king's notice in the traditional way, by being prominent and well connected at court, or by an outstanding record in royal service. They had climbed the accepted political ladder, and got to the top. Such men were unlikely, except under extraordinary pressure, to consider that the system which had served them so well was in need of anything more than superficial reform.

The inertia resulting from the natural disinclination of men in power to question the system by which they ruled was reinforced by the way ministerial government worked. Collective responsibility was unknown, either in doctrine or in practice. Since the only title to power was the favour of the king, the life of a minister was a constant struggle to retain or enhance royal favour, and to limit or diminish the credit of others. Council meetings were the scene of acrimonious clashes as ministers denounced each other's policies in front of the monarch. They also sought to bring more personal pressure to bear at the weekly private sessions in which a minister explained his department's business to the king. Much energy was expended in seeking the favour of women known to be influential with the king, such as Louis XV's mistresses, or Louis XVI's adored queen, Marie Antoinette. Nor were ministers above engineering obstacles to the success of their colleagues' policies after they had gone into effect.

In such a system only resolute conduct by the king himself could prevent intrigue from becoming the main activity of political life. Louis XIV succeeded to some extent by making clear that all ministers could normally rely on his unflinching support so long as they served him honestly and did not question the decisions he made. Even then he was unable to eliminate rivalry between the Colbert and Le Tellier dynasties. Louis XV left everything to a prime minister, Fleury, until 1743. After that he sought to protect his own authority by fomenting rather than discouraging ministerial rivalries. By switching his favour constantly from one faction to another he kept himself out of the pocket of any; but this was at the cost of repeated reversals of policy which inevitably damaged the government's prestige and authority. Under such a style of government, nobody could feel sure that any policy, particularly a radical or reforming one, would be pursued for long; and opponents felt justified in doing all they could to obstruct its success in order to show the king how misplaced his confidence had been.[7] From his accession until 1781, Louis XVI relied on the guidance of a first minister, Maurepas, who selected most of the other ministers and manoeuvred them out of office if they appeared to threaten his primacy. After Maurepas's death, faction once more became the rule, Calonne and his supporters quarrelling constantly with Breteuil and his party. Seeing that Louis was persuaded by Calonne's arguments in 1786, the comptroller-general's rivals hid their reservations about his proposals and promised support. But they did so in the hope that his plans would encounter such difficulties that he would not be able to retain the king's confidence. So even when the state had reached a crisis of such proportions that only sweeping remedies would suffice, several members of the government were more concerned about their own situation than about achieving a satisfactory solution.

Under Louis XVI, moreover, ministerial rivalries did not always end with the dismissal of one or other of the antagonists. When, under his predecessor, ministers fell from power, they were normally disgraced—which meant that they were exiled from the court and capital. When Maurepas was recalled to office in 1774, he had been in exile since 1749. The ministers whom he ousted, Maupeou and Terray, went into enforced exile as was traditional. But after that the practice lapsed, and ex-ministers merely returned to private life without restriction on their activities—as did Turgot and Malesherbes in 1776. Choiseul, disgraced in 1770, profited from the new laxity and returned to Paris, gathered a party, and tried to engineer a recall to power. He failed, but the example did not discourage Necker after he in turn lost power in 1781. Scarcely out of office, he began at once to mobilize public opinion in the hope of forcing himself back into power when the occasion arose. He wrote a three-volume treatise on the *Administration of the Finances* in order to defend his record and propose reforms—incidentally laying bare for the first time many of the state's innermost workings. And he encouraged the formation of a Neckerite party, whose members through writings, salon talk, and political intrigue kept up a constant barrage of criticism against Calonne's policies, and kept the name of the Swiss miracle worker constantly before the eyes of those who were influential in public affairs. Necker was to reap the rewards of this campaign in 1788; but not before it had played its part in exacerbating the difficulties of intervening ministries.[8]

The political and personal hostilities of ministers were often aggravated by inter-conciliar and inter-departmental tensions. The king's council, theoretically a single and indivisible institution, functioned in practice as several different bodies, whose personnel overlapped, but was not identical in every case.[9] The highest decisions of state, mainly those concerned with foreign policy, were taken in the intimacy of the council of state,[10] to which the king summoned a handful of advisers called ministers of state. These ministers might or might not also be secretaries of state, with departmental responsibilities; but in the endless struggle for royal favour, ministers of state had an obvious advantage over their colleagues who were never convoked to the council of state. But even the comptroller-general, who had to find the money to pay for the policies decided on there, had no guaranteed place on this supreme council. Necker resigned in 1781 because his demand for a seat there was refused.[11] In the finance-minister's absence, only the king knew enough of the financial situation to impose caution on the schemes of other ministers, and Necker realized that Louis XVI had neither the presence of mind nor the force of character to do this. In contrast, the heads of all the great departments of state sat on the council of dispatches, where internal policy was decided. This meant the comptroller-general, the chancellor or keeper of the seals (head of the judiciary), and the secretaries of state for foreign affairs, war, the navy, and the royal household; each secretary of state also had administrative responsibility for a number of provinces, on whose affairs he reported to the council.[12] Finally, they all sat, too, on the Privy Council, a sort of supreme administrative tribunal which the king never attended, and which was presided over in his absence by the head of the judiciary. But the Privy Council was not a policy-making body. Its mainstay were the thirty councillors of state, all experienced lawyers or career administrators. Their business was in its turn prepared and presented by the eighty 'masters of requests', rich young men of ambition who had bought their offices and hoped by shining in them to rise further up the governmental ladder. Most ministers not recruited from court circles began their rise to power as masters of requests in the service of the Privy Council, for it was here that they caught the eye of members of the government and established their capacity to be entrusted with important business. As mere masters of requests, they had no power; but if they could acquire a ministerial patron, they could in time come to exercise considerable influence with him and so help formulate the policies he laid before the king.[13]

The immediate ambition of most masters of requests, however, was to become an intendant. The government of the provinces by intendants was the most famous and influential aspect of the French system of government, and the one of which Louis XVI was most proud. Beginning in the early seventeenth century as occasional royal commissioners sent out into the provinces on specialized missions, the intendants had evolved by the middle of Louis XIV's reign into permanent, resident royal agents with wide powers over justice, police, and finance, the direct representatives of the king's authority in each of thirty-four administrative districts (generalities), covering the entire kingdom.[14] Intendants gathered information about their generalities and sent it to the government; they supervised tax collection, public works, measures of public health, and in general all governmental activity. In principle they merely transmitted and carried out royal orders, but in practice they had considerable local initiative. Each generality in its turn was divided into a number of 'subdelegations', headed by subdelegates appointed by the intendant and responsible to him as he was responsible to the central government.

Both contemporaries and subsequent historians have been impressed by the simplicity and apparent efficiency of this structure; and there is no doubt that its personnel was of the highest quality. The eighteenth-century intendants were energetic, conscientious, and imaginative administrators, and most of their subdelegates had a respectable record, too.[15] But energy and ability are not the same thing as power. Historians often quote the famous observations made by John Law in 1720. 'This realm of France is governed by thirty intendants. You have neither parlements, nor committees, nor estates, nor governors. I would almost add, neither king nor ministers; there are thirty masters of requests, posted in the provinces, on whom depends the well-being or misfortune of the provinces, their abundance or their sterility, etc. . . .'[16] But if this had ever been true (which is doubtful), it certainly did not describe the situation under Louis XVI. No intendant was stronger than the support he received from the centre; and with ministers so habitually divided among themselves, ambiguous and contradictory orders often went out to the provinces. No intendant dealt exclusively with a single minister, as his subdelegates dealt with him. Consequently, they usually found themselves reporting to and taking orders from several different ministers at once (depending on the matter in hand), ministers who were often deadly rivals with no interest in the successful implementation of their colleagues' policies. And even when instructions were clear, no intendant could be sure that a change of ministers would not leave him isolated, pursuing a policy that no longer enjoyed governmental support.[17] Nor was an intendant's authority absolute or uniform within his generality. Few generality boundaries were coterminous with military, fiscal, ecclesiastical, or judicial areas, so that the rules and customs which he had to observe might differ widely from one part of his jurisdiction to another. Most intendancies, finally, were chronically understaffed. Ever since they had become permanently established in the provinces, the range of business falling within the competence of the intendants had never ceased to grow. Every new policy was entrusted primarily to them for its execution; every new institution was placed under their surveillance; none of the king's subjects felt justified in taking the slightest public initiative without securing an intendant's sanction. Yet in 1784, the intendant of a province the size of Brittany had only ten clerks and secretaries to assist him.[18] In 1775, ten tax inspectors were expected to check the returns of 3,000 parishes in the generality of Bordeaux.[19] Not only did such understaffing cause constant delays in the dispatch of business, and mean that less urgent matters were badly neglected; it also meant that intendants had to rely for the execution of many of their orders on the compliance of officers and institutions not subject to their direct control. The capacity of such bodies for evasion and resistance was endless.

The intendant was not even the sole representative of the king in any region. The generalities which emerged as administrative units in the seventeenth century superseded but did not replace France's ancient provinces. Nor did the triumph of the intendants bring the abolition of the provincial governors. It is true that the intendants took over many of the governors' former powers, but they did not take over all. Governors or their deputies continued as the supreme military authorities in the thirty-nine (in 1776) *gouvernements*, as the provinces were also known. Without their sanction no intendant could call upon troops, and therefore the governors played an important role in the maintenance of public order. Governors and their deputies were usually princes, dukes, peers, and court noblemen, so they also enjoyed far greater social prestige than the mere masters of requests who occupied the intendancies. Many of them knew the king and his ministers personally, and were able to cultivate these contacts by residence at court. Above all, only governors were deemed persons of sufficient prestige to deal, as the king's representative, with corporate bodies enjoying local power completely independently of the intendant; the provincial estates and the parlements.

The authority of the intendants extended everywhere, but in a third of France they had to share it with local estates. These provinces were known as *pays d'états*, because in them taxes were levied only with the consent of the estates. (Elsewhere they were levied technically by royal magistrates known as *élus,* whose jurisdiction was the *élection*—hence these provinces were called *pays d'élections.*)[20] In the sixteenth century most provinces of France had enjoyed some form of estates representing the three orders of clergy, nobility, and third estate. In the course of the century following, however, many of them had ceased to be convoked as the crown found itself able to establish *élections* in their jurisdictions and so levy taxes without their consent. By the eighteenth century only three major provinces retained their estates—Brittany, Languedoc, and Burgundy—

although Provence enjoyed a diminutive form, the 'Assembly of Communities'. The fifteen minor regions which also still boasted estates had only kept them because of their relative insignificance.[21] But all provincial estates enjoyed the right to bargain over, consent to, and distribute the burden of taxation in their provinces, and when they were not sitting, intermediary commissions carried on their work. By compounding in advance, they were usually able to secure lighter tax loads for their provinces, and at the same time their credit (as representative institutions) was so good that they were often used to borrow money on the government's behalf at rates cheaper than the king himself could command. In the larger *pays d'états* the estates also played a major role in public works such as road building, in poor relief, and in the recruitment and billeting of soldiers. Yet the representative quality of provincial estates was very limited. Hardly any of their members were elected; most were either nominated by the king or sat as of right, being holders of specific ecclesiastical benefices, officers of specific towns or corporations, or (as in Brittany and Burgundy) nobles whose status gave them an automatic seat. Nor did provincial estates usually obstruct the work of governors or intendants (if we except a stormy decade in Brittany in the 1760s).[22] They did, however, complicate the structure of government at the local level, and helped perpetuate many of the privileges, exemptions, and special cases of which Calonne complained with such exasperation in his memorandum of 1786.

The most striking example of this diversity, however, lay in the municipal organization of France. No two towns had quite the same constitution, customs, or privileges; and even when general measures affecting them all were introduced, as in the 1760s, many towns were able to negotiate exemptions for themselves. It is true that most municipalities had three tiers of government—a general assembly, a council (often called *notables*), and an executive (*corps de ville*)—but the names, the size, the composition, or the exact functions of each body were not the same everywhere. The only common feature of any importance was that, ever since 1683, municipal finances and the nomination of officers had been subject to close control by the local intendant.[23] No local taxes might be raised, no debts incurred, and no monies spent without the intendant's prior authorization; and although candidates for municipal office were designated by town councils,[24] the final choice was usually made by the government, on the intendant's advice. This tight administrative control had originally been imposed in the later seventeenth century in order to remedy the chaos into which municipal finances and administration had fallen during a century of political upheaval. By 1760, however, controls seemed to have squeezed all life out of the municipalities, and it was in an attempt to revive them that the comptroller-general, L'Averdy, introduced the municipal reforms of 1764-6.[25] L'Averdy's aim was to reinvigorate municipal life by making most positions of authority elective, by granting towns more financial freedom of action, and by dividing supervision of municipal affairs between the intendants and the courts of law. But no sooner had the difficulties and disruptions which followed these sweeping reforms died down, than they were revoked by the authoritarian Terray in 1771. Terray's successor, Turgot, toyed with further plans for reintroducing municipal elections in 1775, but fell before he could achieve anything; consequently, all the vices which L'Averdy had sought to remedy persisted down to 1789. From the governmental point of view the worst of these was the difficulty of finding able municipal officers who could be relied on to carry out orders. The main duties of such officers—tax collection, police, organization of billeting—won them little popularity, took up much time, and were unpaid. Willing candidates were understandably rare, and those who *were* willing were not always men whom the intendant would ideally prefer. In nominating them, however, he was seldom in a position to disregard the names proposed to him by subdelegates and other local officials.[26]

Governments rely for the most part on the tacit compliance of those subject to them, and in the eighteenth century this was even truer than it is today. To enforce its orders upon subjects who were recalcitrant, the king's overworked, understaffed bureaucracy commanded an even more inadequate number of police. The only police force in the countryside was the *maréchaussée*, a body less than 4,000 strong to cover the whole of France, ill-paid, ill-equipped, and not numerous enough anywhere to confront any but isolated criminals.[27] In towns, policing was the duty of the watch (*guet*), but they were scarcely more numerous, in relation to the numbers of inhabitants, than the *maréchaussée*; for instance, the watch of Lyons, a city of around 150,000 inhabitants, was only 84 strong.[28] Paris, admittedly, was better served; it had a special lieutenant of police who commanded some 3,114 officers of various denominations.[29] But even this was not many for a teeming city of over 600,000. When matters got out of hand, therefore, there was little alternative but to call in the army. Troops were not always on hand when disturbances broke out (although the 3,600-strong *Gardes Françaises* regiment was permanently billeted in Paris and often called upon by the lieutenant of police), and they moved slowly, even when they could be spared from frontier duty. But once deployed, they usually proved decisive in restoring order. All that was required was firmness and determination on the part of the authorities in using them.[30]

Unfortunately, firmness and determination were not qualities frequently displayed by eighteenth-century French governments. In principle, the chain of command from the king downwards was direct, simple, and clear, and generations of historians have been dazzled by these qualities. In practice, vacillation, uncertainty, and inefficiency were built into the system at every level, and probably even kings more self-confident and decisive than Louis XV and Louis XVI could not have eliminated these qualities throughout the administration. Even Louis XIV was increasingly trapped, in his later years, within the machine he had created.[31] This machine had evolved impressive bureaucratic procedures and routines, but it was ill-equipped to implement rapid or frequent changes of policy, and painfully slow to act in emergencies. And these internal weaknesses were only exacerbated by the opposition which government policy was liable to meet at every level—an opposition that came largely from institutions and groups which Louis XIV had preferred to bypass rather than eliminate when building up his governmental structure. While he lived, Louis's disdain for opposition was enough to keep it largely impotent. But when he died, leaving a child on the throne, resentments hitherto held in check burst out, and by the time the government once more felt inclined to restrain them, it was too late. The right to oppose the government had re-established itself, and was not to be denied again.

3. Opposition

Opposition to the government before the Revolution was never focused on a national institution such as the English Parliament. No such institution existed. It is true that medieval and early-modern kings had sometimes convoked the Estates-General, an elective national representative body; but its powers were vague, its composition fluctuating, and its convocation irregular. The last time the Estates-General had met was in 1614, and then its proceedings were more notable for the quarrels between various members than for opposition to the crown.[1] Yet despite its undistinguished history, the tradition of the Estates-General did not die, and in times of crisis those dissatisfied with royal government instinctively turned to the idea of reviving the Estates-General as a means of setting the state to rights. Aristocratic *frondeurs* agitated for it in 1650 and 1651, when it almost met. Those who blamed Louis XIV's rule for the disasters which befell France in the 1690s and 1700s believed that a revived Estates-General would be the best protection against further royal depredations;[2] and this view was also expressed during the post-mortem on Louis's rule which characterized the early years of the Regency.[3] Then in 1771, when Chancellor Maupeou launched a frontal attack on the powers of the parlements, certain of these bodies (previously no friends of the Estates-General, which they saw as a rival) made demands for the ancient national assembly to be called in order to save the state from despotism.[4] Though louder and more public than previous demands, the calls of 1771 were still ignored by the crown. But from then onwards the idea of the Estates never ceased to be discussed and ventilated as a serious political possibility—so much so that Calonne felt obliged to spend time arguing against it before outlining his reform proposals to Louis XVI in 1786.[5]

Provincial estates were no effective substitute for the Estates-General as a forum for opposition. Those which had not been eliminated in the seventeenth century remained in existence precisely because they were more of a help than a hindrance to government.[6] On the rare occasions when provincial estates did prove recalcitrant, as in Brittany during the 1760s, the issues at stake were local rather than national ones; and the sympathy and interest which their struggles evoked elsewhere were nowhere translated into practical measures of support. Municipalities, supine under the domination of the intendants, seldom dared to dispute openly any royal order that came down to them. Popular rebellion, the most formidable and persistent form of opposition that the crown had faced in the seventeenth century, had died out by the eighteenth. The last widespread provincial uprisings had been in 1675, when Brittany and Guienne rose against increasing taxation; the last local outbreak of any importance was the war of the Protestant *Camisard* peasants in the Cevennes between 1702 and 1705. From that time until 1789, nothing more serious than food riots at times of bad harvests challenged the government's authority. Violent incidents by no means disappeared: in Provence, for example, there were on average two outbreaks a year throughout the eighteenth century.[7] Serious and widespread bursts of rioting were also not unknown—such as the famous 'flour war', a panic over bread prices that affected Paris and four northern provinces in April and May 1775.[8] Yet the authorities never had much trouble in restoring order, and in any case the rioters did not see themselves as opposing the government. More often than not they claimed that the king was on their side against the greedy profiteers who were presumed to be keeping bread and grain prices artificially high. It was a pathetic delusion in an age when governments were increasingly inclined to experiment with relaxing age-old controls on the price of foodstuffs; but still the really dangerous opposition to such policies came not from the populace who would suffer most, but from agents of government such as the courts of law who had the unenviable task of keeping public order when decontrolled prices rose.[9] The most effective organs of opposition, in fact, were built into the very machinery of the state itself.

The church, for example, formed a state within the state.

True, the king controlled all appointments to bishoprics, richer abbacies, and other senior benefices, and the days of disputes with the Pope over such matters were long gone. True, too, that the clergy loyally gave spiritual and psychological support to the established order through the educational system which they largely controlled, and by preaching to the populace the virtues of subservience. Yet the church remained an independent corporation, with its own undisclosed income in the form of tithes and revenues from estates covering perhaps a tenth of the land of France. The clergy were exempt from direct taxation, paying instead only a regular 'free gift' (*don gratuit*) to the king. This gift was made after negotiations between the crown and the elected Assembly of the Clergy, which met every five years,[10] the only national representative body of any sort to meet regularly in France throughout the seventeenth and eighteenth centuries. The chief reason why this apparent anomaly had survived was that, like the provincial estates of Brittany or Languedoc, the Assembly of the Clergy could borrow money at rates more advantageous than the government (4 to 5 per cent rather than 8 per cent), using the church's enormous landed wealth as security. Throughout the eighteenth century, the whole sum of each 'free gift' was always borrowed, the interest being paid by self-imposed clerical income taxes (*décimes*).[11] As a result, by the mid-1780s the capital debt of the clergy had risen to some 140 million *livres*,[12] a heavy burden no doubt, but a formidable guarantee against governmental interference with the church's institutional independence. Not that the government was always deterred by such difficulties, especially when the expenses of war drove it to explore hitherto untapped sources of revenue. Thus is 1749, the comptroller-general, Machault, in introducing his new *vingtième* tax, declared that the clergy, like all other privileged groups, should not be exempt from it, and that to facilitate its accurate assessment a public survey of the value of ecclesiastical property should be made. These proposals provoked a two-year crisis in which the church showed what formidable opposition it could put up.[13] The Assembly of the Clergy of 1750 denounced Machault's plans and refused all co-operation with him. Most of the clergy took the same stance even after the Assembly had been dissolved, and eventually, in December 1751, the government abandoned all its proposals. In 1780, a renewed attempt was made to value church property for taxation purposes when it was announced that, as from 1785, ecclesiastical landlords would be required to make declarations of value (*foi et hommage*) to the king as their feudal overlord; these declarations had first been required in 1726, but repeated stays of execution had hitherto been granted. Once more the clergy organized a campaign of resistance in the face of which, in 1785, the government withdrew.[14] The fact is that the clergy were, with the possible exception of the armed forces, the best-organized and best-disciplined body in the kingdom, and while they were normally happy to place these accomplishments at the disposal of the king to deploy as he thought fit, there was a price to pay. The crown must respect the church's privileges—its autonomy, its right not to reveal its wealth, and its tax-exemption. When policies appeared to threaten these, the church would go to extreme lengths of non-co-operation, public agitation, and even political blackmail when negotiating 'free gifts', in order to defend itself. And the leading prelates who managed the church's affairs were strengthened by their unrivalled political experience in managing assemblies and negotiating with the crown. Few ministers were determined and resourceful enough to risk an open confrontation with such an adversary; and those who were knew that, like Machault, they might be abandoned at any time by an irresolute king who hated trouble.

Yet at least the clergy asked nothing more than to be left alone, and normally opposed only direct attacks on themselves. By contrast, the parlements claimed a right to interfere and oppose the government over its whole range of activity. It was from them that the crown encountered most formidable and most persistent opposition throughout the eighteenth century.

The parlements were 'sovereign' courts of law, final courts of appeal for the thirteen (very unequal) judicial districts into which most of France was divided.[15] There were also twenty-one other sovereign courts, most of them specializing in fiscal affairs, whose outlook, traditions, and fields of recruitment were similar to those of the parlements.[16] The magistrates of these sovereign courts, numbering in all around 2,300 in the late eighteenth century, constituted the famous *noblesse de robe*—so-called because the offices they occupied conferred nobility.[17] They were the kingdom's governing élite, for not only did they staff the higher judiciary, by definition; almost all intendants and councillors of state and a majority of the king's ministers were also recruited, directly or indirectly, from the ranks of those who had passed through the sovereign courts.[18] But those who remained in the courts had a major advantage over those who went on to try their ambitions in government; judges were irremoveable. In the course of the sixteenth and seventeenth centuries a government hard pressed for money had turned increasingly to the sale of offices; and those in the sovereign courts, with the privileges and the ennoblement that they conferred, were among the most easily saleable. And so public office had become private property, and the payment of certain taxes ensured that an office could also be bequeathed to the holder's heirs. This system, brilliant as a means of raising money in the past, tied the government's hands almost irrevocably for the future, because it meant that a magistrate could only be dismissed from his office if the government could refund the purchasing price. Few ministries, lurching from one financial crisis to another, were in a position to do this; and when a large-scale suppression of higher judicial offices was attempted between 1771 and 1774, the promised compensation had to be staggered over up to twenty years.[19] It took the Revolution to abolish venality of offices, and even then it could only be done by granting compensation on extremely arbitrary terms.[20] Before 1789, magistrates could be confident that whatever they said or did, the king would not dismiss them.

The independence of the parlements' members gave added strength to a long tradition of opposition, for the sovereign courts had a recognized constitutional right to criticize royal policies through remonstrances. No law took effect in France until it had been 'registered' by the courts that had to enforce it; and none would be operative throughout France until it had been registered by every one of the parlements and sovereign councils for their own areas. But before registration, a sovereign court had the right to point out defects and disadvantages in new laws through remonstrances sent privately to the king. If the king chose to ignore a remonstrance, the court was obliged to obey, but it was easy to delay registration pending further remonstrances, and during that time the law in question remained inoperative throughout the court's jurisdiction. Alternatively, the law could be registered with modifications. The king could then order registration of the original, purely and simply, in *lettres de jussion*, but if this had no effect he would go in person to the court (in Paris; in the provinces the governor would represent him) and conduct a forced registration (*lit de justice*). This was the last word. Parlements might, and often did, denounce forced registrations afterwards and declare them illegal, but they seldom resisted a law once it was on their registers by the king's express command.

This did not, however, exhaust the parlements' capacity for opposing the government and its agents. Not only did they, as courts of law, hear and decide cases brought before them. They also had the right to make regulations and bylaws (*arrêts de règlement*) for their own areas, and suspend the effect of general laws in case of necessity. In this capacity they often fixed prices, censored books, generally forbade practices which they deemed undesirable, and interfered in municipal government. In all these fields they came into direct rivalry with the intendants. It would be wrong to exaggerate this rivalry. Often intendants invoked the help of the parlements to reinforce their authority, and co-operation between the two in maintaining public order was the normal state of affairs. But some serious and spectacular clashes of authority were inevitable, and intendants could not always be sure of emerging triumphant. The parlement of Bordeaux, for example, effectively drove out two intendants in the later eighteenth century when ministers became convinced that the court had a case against them, and recalled them to Paris.[21]

In other words the government wavered. For, despite the independent tenure of magistrates, despite the rights of registration and remonstrance, and the power to issue administrative regulations, the parlements were only strong when the government was weak. The experience of the Fronde, when the resistance of the parlement of Paris had paralysed a divided and faction-ridden regency government, left an indelible impression on the young Louis XIV, and he was determined to brook no trouble from the courts after he assumed personal power in 1661.[22] In 1673, he decreed that henceforth registration of new laws must precede remonstrance, and this effectively killed remonstrances as a vehicle for opposition for the rest of the reign. It did not (contrary to the claims of traditional historiography) leave the parlements in docile silence until 1715; there remained many ways in which government policy could still be resisted, twisted, and evaded in its execution by the courts.[23] But open opposition largely lapsed between 1675 and 1714, for the parlements knew that they risked swift and brutal counter-measures—such as the long exiles suffered by those of Rennes and Bordeaux after the uprisings of 1675.

In 1715, however, the situation changed. When Louis died, leaving his nephew Orléans as regent for his five-year-old heir, the regent needed the sanction of the parlement of Paris in order to overturn clauses in the old king's will which circumscribed his power. In return he revoked the law of 1673, so that the parlements were once more free to remonstrate before registering laws. The result was a renewed period of opposition down to 1732, when the parlement of Paris denounced, in remonstrances that were increasingly often printed and publicly distributed, government financial measures like John Law's 'system', and attempts to silence Jansenist critics of the church by making the Bull *Unigenitus* of 1713 a law of the state.[24] During this period the weapons for future conflicts were forged: on the parlement's side repeated remonstrances, modified registrations, and judicial strikes; on the government's, *lits de justice*, arrests of leading magistrates, and exiles of the parlement to small and uncomfortable provincial towns. But the root cause of these clashes was the instability of governments during Louis XV's minority. Once Cardinal Fleury was fully in control, by around 1730, government policy became once more firm and consistent, *Unigenitus* became a law of the state, and all the parlement's protests proved unavailing.

The high point of the parlements' power came between 1749 and 1771, a time of acute financial difficulty for the state owing to the costs of warfare, and great ministerial instability resulting from the factionalism which Louis XV fostered among his ministers after Fleury's death. No policy seemed able to command the unanimous support of ministers, and those who opposed any particular measure naturally took the side of the parlements if they attacked it. No resolute action to silence them could be expected when every minister knew that the parlements could be invaluable allies in struggles with rivals.[25] The chief feature of this new phase of activity was the emergence of the provincial courts as pace-setters of opposition. The parlement of Paris continued to be preoccupied with religion, defending the remnants of the Jansenist party in the church against the persecution of the orthodox. In 1760, a well-organized clique of Jansenist magistrates engineered an attack on the greatest advocates of orthodoxy, the Jesuits. The other parlements rapidly joined the hunt, a number of ministers disliked or were indifferent to the Jesuits, and so the government was manoeuvred in 1762 into dissolving the Society of Jesus in France entirely.[26] But on the question of the new *vingtième* tax introduced in 1749, its doubling in 1756, and its trebling in 1760, the parlement of Paris issued only muted protests which were easily overridden. Real resistance came from the provincial sovereign courts, who sent in repeated remonstrances, deferred registration, and obstructed government agents who attempted to gather the new levies. Intendants proved unable, single-handed, to overcome recalcitrant parlements, and forced registrations conducted by military governors became common in the provinces.[27] When the government began, in 1755, to experiment with elevating a single sovereign court, the *Grand Conseil*, above all others for the registration of new laws, certain parlements responded by declaring that they were all parts, or *classes* as they put it, of a single national court. The implication was that in future they would act together.[28]

Yet apart from the harrying of the Jesuits, nothing approaching united action ever materialized, for the parlements were deeply divided both within themselves and among themselves about almost everything–including the theory of *classes* itself.[29] Even when their action appeared unanimous, as in 1763 when the government tried to introduce a land survey in order to make the levying of the *vingtièmes* more accurate, there was little co-ordination between their various protests and only a few courts persisted long in their opposition. Once again the government's withdrawal of the plan owed as much to a divided ministry as to the strength of the parlements.[30] When, in 1763 and 1764, restrictions on the grain trade were relaxed, some parlements denounced the measures, while others applauded them.[31] Even when the government, now adopting a harder line, remodelled the parlements of Pau and Rennes in 1765, not all parlements espoused the cause of their purged brethren as instantly as might have been expected; the issues were primarily local ones which could not sustain the attention of courts in other provinces, which had their own problems.[32]

But the Brittany affair was mismanaged. The aggrieved governor of Brittany, the duc d'Aiguillon, was allowed to lay his record before the parlement of Paris; then, when the magistrates began to probe into government secrets, the case was closed by royal order. The protests which naturally followed were met with deliberate provocation by a new chancellor, Maupeou. He was hoping to use a confrontation with the parlement of Paris to discredit and undermine the position of the duc de Choiseul, who had dominated the government since 1758.[33] In this he succeeded; but when, having secured Choiseul's dismissal at the end of 1770, he tried to rebuild his relations with the Paris parlement, mutual suspicion proved too much, and he was forced to unforeseen extremes. The magistrates were exiled, and the parlement's membership and powers completely remodelled. When the provincial courts protested, they were remodelled too. By the end of 1771, the higher judiciary of France had been completely transformed. Its numbers had been halved, its competence diminished by the elevation of new subordinate courts (*conseils supérieurs*), its old system of fees replaced by free justice and salaries for magistrates; and above all, venality of office, and the security of tenure that went with it, had been abolished.

Maupeou's new order, which lasted until Louis XV's death in 1774, remains controversial among historians.[34] But all sides agree on one thing—Maupeou did reduce the parlements to relative silence. Despite defiant protestations throughout 1771, the government had no trouble in dissolving the old parlements, and even persuaded large numbers of magistrates to co-operate in the new courts. They were no more united during this supreme crisis than they had been during the apparently more successful days of the 1750s and 60s. Maupeou, on the other hand, showed what a firm and determined government could do. He did not even tamper with the rights of registration and remonstrance: he did not need to. The main source of opposition to the government had been masterfully tamed. An opportunity had been created to carry through far-reaching reforms with little or no resistance.

But little advantage was taken of it. Apart from the further prolongation of the *vingtièmes* at the end of 1771, and the resumption of the task of reassessing liability for them which had been suspended since 1763, nothing done during the three and a half years of the new order would have provoked much trouble from the old parlements. And this was because none of it was very radical. Maupeou talked of recodifying the law, but did nothing. Louis XV himself vetoed further changes in the judicial structure. Outside the parlements, venality of office remained untouched. The efforts of Terray to diminish the role of private businessmen in the management of royal finances,[35] perhaps the most pregnant innovation of his ministry, would have won the applause of magistrates always deeply suspicious of what they called 'capitalists' —financiers who made fat profits from lending their money and services to the government.[36] A Necker or a Calonne might have made more of the opportunity created by Maupeou, but the ministry which blundered into the reform of 1771 saw it as an end in itself, rather than a door to more far-reaching changes.

In any case, when Louis XV died, Maupeou was dismissed and the reforms were abandoned. Even while Louis lived, their finality was never beyond question,[37] and Louis XVI and Maurepas, who guided all his steps until 1781, wished to begin the new reign by a spectacular break with the legacy of the old king. The new comptroller-general, Turgot, a serious reformer, thought he could carry out his programme without obstruction by the parlements, and advised the restoration of the old judicial order. And he was right. The protests of the parlement of Paris against his proposals were swept aside in a *lit de justice*. His programme was only abandoned, some months later, after he had fallen victim to court intrigue. The restoration of the parlements in 1774 has often been condemned by historians as the monarchy's fatal mistake,[38] on the grounds that they blocked all subsequent attempts at reform. But in reality they blocked nothing of national importance. The government, at least until 1783, was firm and relatively united, and the tax rises of the early 1780s met little serious resistance. The parlement of Paris, chastened by the Maupeou experience and reluctant to risk a repetition, was more docile between 1776 and 1786 than it had been since the days of Fleury. The provincial courts were indeed more turbulent, and on a local level they scored some notable victories over intendants; but several of them were paralysed for years by personal recriminations between magistrates who had co-operated with Maupeou and those who had not,[39] and they remained deeply divided over the great issues of the day. The parlements had, in fact, passed the peak of their power by the reign of Louis XVI. If, in the 1760s, they had thought themselves indispensable, Maupeou had disabused them. After 1774, while they trod more warily, the government dealt with them much less nervously, knowing how easily they could be curbed. Whether, despite all this, they would have united to block broader plans of reform, we cannot say. Much of Turgot's programme never reached them, and he expected in any case that their resistance could be overcome. Necker was less sanguine, and pondered ways of circumventing them.[40] But the point is that, for whatever reasons, until 1786 no comprehensive reform plans appeared. Most ministers did not see the necessity, and Turgot and Necker, the only two who did, were destroyed by the intrigues of their rivals before they could unveil more than a fraction of their programmes.

Opposition to the government in eighteenth-century France

was persistent, loud, and often spectacular. But if ministers chose to ignore or override it, its fundamental weakness soon became apparent. It was not the strength of opposition that prevented the crown from reforming, but the inertia, uncertainty, and irresolution of the crown itself.

Why then was the government so often unsure of itself? The irresolution of kings and the rivalries of ministers, personal factors, obviously played a far larger part than many historians are prepared to admit. But the problem was more complex than this. This was the Age of Enlightenment, a time when hallowed orthodoxies were being questioned and publicly doubted, while every year saw the appearance of new panaceas for the world's problems. Few writers had presumed to tell the government of Louis XIV what it should do, or how it should govern the country; and Louis and his ministers certainly wasted no time in listening to those who did. In the eighteenth century, theorists of every persuasion made it their business to press policies upon the state, and for support, they appealed to the force of informed public opinion. The monarchs themselves, preferring to hunt rather than read, absorbed little of this propaganda. But their cultivated, well-informed ministers took it seriously, and from time to time tried to put new, untried doctrines into practice. These experiments, their uncertain results, and the criticism they aroused, made eighteenth-century governments even more volatile. To understand the breakdown of the Old Regime, therefore, we must understand the climate of ideas in which it tried to operate.

4. Public Opinion

The one class whose rise no historian disputes in the eighteenth century is that of the educated general reader. The evidence for its rise does not lie in the statistics of literacy, although here, too, there was a rise. On the eve of the Revolution perhaps 63 per cent of the French population could neither read nor write, whereas a century beforehand the proportion had been 79 per cent.[1] This improvement certainly owed a good deal to better educational provisions, but most of the technically literate could not really be called educated. When they chose to read at all their preferred fare was one of almanacks, chapbooks, and cheaply produced collections of traditional stories and tales of wonder.[2] Popular literature provided an escape from the harsh realities of the everyday world rather than a key to exploring its complexities. It confirmed familiar horizons rather than opening up new ones.

The rise of the educated general reader is, however, abundantly clear from other evidence. There was, for instance, a marked expansion in the book trade. The sources, however fragmentary and incomplete, all point to a steady expansion of book production from the beginning of the century down to the early 1770s. And from then on they show a veritable explosion which reached its climax, not to be surpassed again for twenty-five years, in 1788.[3] Another indication is the growth in the number, size, and frequency of newspapers and journals. At the beginning of the century the French periodical press consisted of three semi-official journals, closely supervised by the government, and a number of French-language periodicals published beyond the reach of Louis XIV's censorship in the Dutch republic. By 1765, the number of periodicals published in France had risen to nineteen,[4] with some of the more famous and informative journals yet to appear over the next decade. In 1777, a daily newspaper made its appearance in Paris; in 1748 the first weekly provincial paper began to be produced in Lyons; and by the 1780s most major (and some minor) provincial towns had their own papers, too.[5] The very survival and prosperity of all these publications, despite often spectacular vicissitudes, shows that there was a growing market for what they offered. 'Gone are the days', declared the *Journal encyclopédique* in 1758,[6] 'when journals were only for the learned . . . Nowadays, everybody reads or wants to read about everything.'

'Everybody' obviously did not mean all those who could read. Books and journals were far from cheap, and although, thanks to improvements in productive and distributive techniques, they cost less than they had in the previous century, they were still beyond the reach of all but the most comfortably off. The price of a single issue (carriage paid) of a major journal in the 1760s would be twenty-four *livres* in Paris and as much as thirty-three *livres* in the provinces[7] — around twice the weekly wage of a skilled workman. The first edition of Diderot's great *Encyclopédie* cost the equivalent of ninety-three weeks' wages, and even the cheapest subsequent edition would have taken fifteen and a half weeks.[8] In any case, only those who had had a long and expensive education were likely to appreciate the contents of such works. But such people were now more numerous than ever before, and it became increasingly easy for them to join or subscribe to institutions which bought all the latest journals, and many of the latest books, too, and provided premises where they could be read.

At the highest and most exclusive level there were provincial academies, of which there were nine in 1710, twenty-four by 1750, and at least thirty-five by 1789.[9] Most of them had libraries or reading rooms, and sponsored prize essay competitions on topics of current interest. Private literary societies also proliferated. A rough count reveals between forty and fifty from mid-century to 1789,[10] calling themselves literary societies, patriotic societies, *musées, lycées, logopanthées*, and so on. They, too, often had libraries and reading rooms, organized public lectures, discussed matters of current concern, and, unlike academies, were relatively easy to get into.[11] Those deterred even by this could simply join one of the many public libraries or reading rooms, where anybody on payment of a fee could have access to all the latest literary productions. For instance, at M. Hubert's 'literary chamber', which opened in Bourges in 1785, one could read all the major journals for an annual subscription of twenty-four *livres*, less than the cost of a single issue of one of them.[12] Such developments put the wider world of the arts, ideas, and politics within the reach of any literate provincial with the means and the inclination to spend a modest sum on informing himself; and the ranks of those who were prepared to do this transcended the traditional social divisions. More and more, nobles and bourgeois alike came to share the same tastes, and to discuss the same issues, in commonly agreed terms. The public life of eighteenth-century France was acted out before an expanding, nation-wide audience of informed public opinion.

In the seventeenth century public opinion had been something the government was determined to control. Louis XIV's ministers exercised minute supervision over everything that was either printed or published in France. Royal censors prevented the publication of anything that seemed likely to provoke controversy and awaken public doubts about the established order. Unsupervised discussion of religious or public affairs either circulated in manuscript or was published abroad, largely in the Dutch republic.[13] But almost as soon as Louis died, the vigilance of the government began to flag. The Regency witnessed a great outburst of writings that would formerly have been instantly suppressed, and although censorship became active again from 1723, it proved impossible to reimpose the iron control of Louis XIV's time. The practice began to grow of granting 'tacit permission' for the publication of works that the government did not object to but felt it imprudent to sanction publicly. In the hands of Malesherbes, the broadminded magistrate who was the chief government censor between 1750 and 1763, this double standard was applied to allow a whole range of publications that would have scandalized his predecessors.[14] It was mainly thanks to his protection that Diderot's *Encyclopédie*, that great compendium of advanced opinions, survived the storms of its early years. 'It is not', Malesherbes wrote, 'in rigour that a remedy should be sought, but in tolerance. The trade in books today is too widespread and the public is too avid for them for it to be possible to constrain it . . . on a taste which has become dominant. So I only know one way to enforce prohibitions, and that is to issue very few of them. They will only be respected when they are rare.'[15] His successors did not share these principles. Several of them believed that the latitude allowed by Malesherbes had opened the floodgates to dangerous and subversive opinions, and that only vigorous action could restore order. Under Maupeou, accordingly, there was a return to severity, and Terray struck at the boldest literature by imposing heavy duties on imported books. The reign of Louis XVI brought no relaxation in this respect, for the king himself was outraged by the irreverent writings sold throughout Paris. 'I can see that it is difficult for you to manage to discover the authors of songs and pamphlets', he told the police lieutenant of Paris in 1782,[16] 'but try to stop these vicious things by whatever means you can.' The next year Vergennes, obsessed like his master by the dangers of subversive literature, imposed stringent new controls on book imports, which left the book trade in chaos right down to 1789.[17] Throughout the century writers were also harassed by the random persecutions of the parlements, although books publicly burned by these bodies often sold better as a result of the notoriety it gave them. But by 1770 the damage, if damage it was, had been done. Even if it had been possible after that date to suppress all contentious literature appearing in France, and to dam up the multifarious channels by which it flowed in from abroad, there was no recalling what had been published in previous decades, and no blunting of the public's appetite for more. The government had lost control of public opinion.

Not only is this shown by the way controversial writings continued to find publishers and buyers. It is also clear from the way the censoring authorities themselves increasingly wooed the good opinion of the reading public. Among the first in the field were the parlements, who had their own printing presses and used them to disseminate innumerable copies of judgements, resolutions, and above all remonstrances, even though the last were strictly supposed to be private communications between the king and his judges.

The parlements were deeply concerned about their public standing. They knew that public support was one of the keys to their power; and massive evidence, throughout the century, that this support really existed showed that the public valued them as the sole bodies empowered to voice grievances.[18] The government, too, grew increasingly interested in maintaining a good public reputation. It sought to monitor the mood of Paris by an elaborate network of spies,[19] and ministers received daily reports from the police lieutenant.[20] Above all, they went into ever more elaborate public justifications for all they did. The rejection of remonstrances, or the quashing of acts of judicial defiance, proved occasions for long printed rebuttals of the sovereign courts' pretensions. When Maupeou attacked the parlements in 1771, he not only took trouble to dress up his action as a preliminary to long-needed reforms in the legal system, he also engaged a team of hack writers to rebut the protests of pro-parlementaire pamphleteers and trumpet the merits of the new order.[21] Turgot's famous edicts to liberalize French economic life, five years later, all had lengthy preambles in which the minister tried to convince the public of the value of what he was doing; and in the provinces even intendants took up the pen to extol the virtues of their own policies.[22] Necker took the public into his confidence to an unprecedented degree when he published the royal accounts for the first time in his *Compte rendu* of 1781. It was a huge success; and although it failed to increase his standing in the government as he had hoped, even after his fall he continued to exploit public opinion for his own political purposes by copious writings in defence of his record in office. And like the parlements, the government was clearly influenced by what it took public opinion to be. Louis XV vetoed Maupeou's proposal to suppress all the provincial parlements completely in 1771 on the grounds that it would make him appear a despot.[23] When, in 1774, a new king restored the old parlements and abandoned Maupeou and his system, he did so largely because this was what he thought the public expected of him.[24]

By the 1770s, in other words, public opinion had become a powerful political force, limiting the government's freedom of action in one sphere, pushing it in unpremeditated directions in another. 'This monarchy', observed the Swedish ambassador in July 1786, 'only differs from despotism through the influence of public opinion. It is the citizen's sole safeguard.'[25] It was also an unpredictable influence—shifting, capricious, an easy prey to transient fashions—which did nothing to increase the stability or fixity of purpose of the king's government.

What, then, were the main characteristics of public opinion in the 1780s? It is no easier for historians than for contemporaries to be sure. It does, however, seem increasingly certain that the public was far from won over by the most sustained assault to which it had been subjected over the century, that of the Enlightenment. This is no place to attempt a definition of that movement, but nothing would have surprised contemporaries more than the tendency of some recent historians to identify it with the whole of eighteenth-century thought. Those who thought themselves 'enlightened' believed they were a small band of crusaders against widespread ways of thinking, habits, and institutions that were not. They believed that the past had bequeathed a huge mass of ignorance, prejudice, and superstition which enlightened men had a duty to work to eradicate. The means of doing this was the propagation and popularization of rational and empirical principles of enquiry first elaborated in the late seventeenth century. The self-styled 'philosophers' attempted to discuss complex and difficult ideas in terms which any person of moderate cultivation might understand. They made a deliberate attempt, in fact, to enlist the sympathies of the new reading public of the eighteenth century. They could not have done so if the authorities had remained determined to restrict all free discussion of sensitive issues; for *philosophes* were interested in fundamental questions. The Enlightenment was a critical movement, and its advocates spent much of their time pouring scorn on pillars of the established order of things, such as the church and the legal system. But, as the century progressed, some at least of the highly educated, well-read men who governed France evidently became persuaded that freer discussion of such matters was in the public interest, and constituted no real danger to the established order.

Nor should we rush to the conclusion—in view of what later happened—that they were entirely wrong. There is no doubt that, by scouring the writings of the *philosophes*, a revolutionary ideology could be pieced together. Those who later blamed the Revolution on the Enlightenment did just this. But there is equally little doubt that nobody before 1789 thought of such a thing. Hardly anybody, and certainly not the *philosophes*, dreamed of revolution, or would even have understood the idea. Certainly they advocated a change in outlook, a way of looking at the world less dependent upon religion and tradition. They urged men to judge things not according to their familiarity, or the dictates of authority, but rather according to whether they were useful to humanity and functioned reasonably. But no conscientious administrator could have objected to the spread of such standards. They encouraged a readier public acceptance of necessary reforms. It was only when the established order had collapsed completely, and when it became obvious that institutions must be recast afresh from their very foundations, that the attitudes propagated by the Enlightenment were to lead Frenchmen in really new, uncharted—in fact, revolutionary—directions.

So dazzled have historians been by the range, eloquence, and verve of the *philosophes*' writings, that they have sometimes too readily assumed that they carried all before them. It is true that by the 1770s they had conquered the *salons*, the academies, and all the established institutions of intellectual life,[26] but their impact on the public at large is much less clear. Recently, influential opponents of the Enlightenment have begun to attract attention—men like Fréron, or Linguet, or the widely read Jesuit journal, the *Mémoires de Trévoux*.[27] Recently, too, doubts have arisen as to whether reforms applauded by *philosophes* owed anything to their influence in practice.[28] Excessive attention has surely been paid to the extremely small, rarefied world of Parisian *salon* society, where new ideas came and went like the latest fashions, but had little echo in the provinces. Over 60 per cent of the books sold in provincial France during the 1780s, for instance, were still religious works,[29] whilst among theological books authorized by royal censors there was a growth in anti-philosophic polemics,[30] suggesting a willingness among defenders of orthodoxy to meet or rebut the attacks of hostile critics. Even in the excited atmosphere of the spring of 1789, the evidence of the *cahiers* suggests that the minds of literate Frenchmen were deeply confused about what changes in politics and society were desirable, if indeed great changes were desirable at all.[31] Quite evidently the French reading public was far from radicalized by the Enlightenment. Indeed, the more radical the idea, the less likely it was to receive acceptance, as many an aspiring writer found to his cost.[32]

Yet it would be equally rash to deny the Enlightenment any influence. The works of Voltaire obviously enjoyed an enormous circulation. The *Encyclopédie* penetrated the country in its various editions, sold well in provincial capitals, and must have been well known to large sections of the reading public.[33] The same could be said of many other 'philosophic' works. Anyone who reads the remonstrances of the parlements in the later eighteenth century will find them steeped (consciously or otherwise) in the ideas of Montesquieu,[34] while political pamphlets from the 1770s onwards are full of the notions and language of Rousseau's *Social Contract.*[35] And whether or not it resulted from their anti-clerical preachings, there was a marked decline in religious commitment in the upper reaches of society which *philosophes* could only welcome. New publications on religious and theological subjects shrank steadily.[36] In Provence (a much-studied area recently) between mid-century and 1789 the volume of pious bequests by testators fell by a half,[37] while sociable men forsook the fraternities of penitents which they had traditionally patronized in order to join masonic lodges.[38] Indeed the spread of Freemasonry, one of the most striking social developments of the century, suggests that the leisured classes who were its mainstay had inner needs which the church was no longer able to meet.[39]

Anti-clericalism was certainly one of the more obvious features of public opinion. Any attack on the position or privileges of the established church was assured of widespread support. It was evident in the public's indifference to the expulsion of the Jesuits in the early 1760s—although the driving force behind this expulsion was neither popular nor philosophic anti-clericalism but rather the traditional hostility to the Jesuits of pious Jansenists in the parlement of Paris.[40] But the half-century of dogmatic wranglings and ruthless feuding among opposed clerical factions, of which the expulsion of the Jesuits was the climax, had brought the clergy as a whole into contempt far more effectively than any philosophic propaganda.[41] Hardly anybody, not even the *philosophes*, wished to destroy the church; but an increasingly worldly, utilitarian public was ready to applaud almost any attack upon the way it was organized and the way it functioned, in the belief that it needed comprehensive reform to make it more responsible to the nation. So there was widespread approval when in 1766 the church set up a commission to dissolve small and underinhabited monasteries and put their revenues to better use.[42] Another popular move was the raising in 1768 of the *portion congrue*, the stipend allowed by monastic impropriators to priests whose livings they owned. The only criticism heard was that the rise was not enough. Clerical meddling in politics, on the other hand, was deeply unpopular. It made Christophe de Beaumont, archbishop of Paris between 1746 and 1781, the most hated prelate of the century, at least before Loménie de Brienne. And one of the reasons for Maupeou's unpopularity was the widespread belief that he was a puppet of the banished Jesuits and was working to reinstate them. Such an atmosphere of public disquiet and suspicion towards the church led ministers like Calonne, not unreasonably, to feel that the time was perhaps ripe for clipping its wings.

There was also growing public support for a variety of social and institutional reforms. Against violent clerical opposition, belief in religious toleration steadily gained ground. The most celebrated case of religious prejudice of the century, the execution of the Protestant Jean Calas in Toulouse in 1762 for the alleged murder of his Catholically inclined son, made news precisely because such sectarian bigotry was already quite uncommon.[43] By the 1780s, several leading archbishops had become advocates of toleration, and even if the bulk of the lower clergy still held out against any suggestion of it, they were increasingly isolated from the educated laity.[44] Cases like that of Calas, brilliantly exploited by the journalism of Voltaire, also raised the question of reform of the law. Maupeou was not interested in such things, but he sought to make his attack on the parlements more popular by declaring it a preliminary to far-reaching legal reforms. The 1760s witnessed a great outburst of writings on the defects of the criminal law, and the campaign was encouraged by the abolition of certain forms of torture in 1780. In 1786 and 1787, a series of cases of alleged miscarriage of justice, brought to public attention by a new generation of publicists who had learned from Voltaire's example, aroused more interest than anything except the Assembly of Notables itself.[45]

But how was reform to be achieved? Voltaire, its most famous advocate, was indifferent to the means. He would support any regime which seemed favourable to it, and hence his support for Maupeou.[46] The public, however, was more guarded. Better no reform, most people thought, if it could be accomplished only by despotism. The idea of despotism as the eighteenth century understood it had first emerged among critics of Louis XIV's arbitrary, unrestrained method of government. But it was given its classic definition in 1748 by Montesquieu in his *Esprit des Lois.*[47] Monarchy, said Montesquieu, was the government of one man according to law. Despotism, the worst of all possible governments, was the government of one man according to no law but his own caprices. Despotism was essentially arbitrary. It had no limitations, and under despotic rule no man had anything that he might call his own. The *Esprit des Lois*, sprawling, curious, and unbalanced as it was, was to prove one of the most influential works of the eighteenth century,[48] and it defined despotism in a way that few later writers chose to dispute. Montesquieu had been a president in the parlement of Bordeaux, and in his theory he had reserved a special place for the sovereign courts. They, along with the nobility, were essential 'intermediary bodies' which in a monarchy stood between the ruler and his subjects. By their power to resist him they prevented him from degenerating into a despot. It is hardly surprising to find the parlements adopting this role with enthusiasm in their disputes with the king in the second half of the century. They portrayed themselves as the sole bulwark against the crown's persistent despotic tendencies, and there is plenty of evidence that, down to 1788 at least, public opinion largely accepted this portrayal. But neither the parlements' concept of despotism, nor that of the public at large, was too closely circumscribed by Montesquieu's precise definitions. Already by the 1760s it was a term used for any form of arbitrary power, or indeed any species of power or authority that the user did not like.[49] Thus Jansenists repeatedly accused Jesuits of despotism; they in turn were said by *philosophes* to nourish despotic ambitions to smother the free expression of opinions. Parish clergy called their bishops despots; taxpayers denounced tax-collectors in the same terms; even spurned lovers found the objects of their desire despotic. Above all, however, despotism was a charge hurled with ever-increasing frequency at the government and its agents. By the 1780s it almost seemed as if government and despotism were synonyms in the public mind. And this suggests that the old order had lost the confidence of those who lived under it.

The first shocks to public confidence doubtless came from France's shattering defeats at the hands of Great Britain in the Seven Years War. But this damage could be undone, as the glorious successes of the American war showed in the late 1770s. More serious were the attempts made in the 1760s to liberalize the grain trade. These measures were perhaps the most outstanding achievement of Enlightenment propaganda. Their adoption by the government represented the triumph of a small but vocal group of economic theorists, the physiocrats, who from the late 1750s had been advocating the abolition of what they regarded as artificial obstacles in the way of the 'natural' economic order of an unimpeded market for the products of the land.[50] But these so-called 'artificial' obstacles were made up of an elaborate network of controls by which the king's government attempted to ensure that his subjects had adequate supplies of bread at prices they could afford. For the king to abolish them was for him to wash his hands of this time-honoured obligation, one of his oldest and most important responsibilities. Accordingly, the first attempts to relax the regulations in 1763 and 1764 were greeted by a wave of protest from the various agencies of public order, who felt betrayed and unsure whether they would be able to maintain public tranquility in the event of bad harvests and high prices.[51] These fears proved well justified when leaner times arrived; there were bread riots and popular panics, and stern suppression of them did nothing to reconcile the hungry populace to a government that had seemingly stopped protecting it.[52] No wonder wide sections of the population were prepared to believe, in later years, that Louis XV had entered into a 'famine pact' with unscrupulous speculators to starve his own subjects.[53] The abandonment of long-established practices, of course, suggested that ministers, too, were losing confidence in traditional policies; and although controls were reimposed in the early 1770s, they were again abolished by Turgot, leaving an over-all impression that the government no longer knew for sure what it wanted. And finally, the arbitrary abolition of a whole complex, well-established apparatus of controls laid the government open to further charges of despotism. Here were measures of incalculable importance being introduced, without any prior consultation, by a handful of ministers apparently on the persuasion of a handful of armchair theorists.

If the 1760s saw public confidence in the way France was governed begin to waver, the 1770s saw it collapse almost completely. The turning point was the ministry of Maupeou and Terray. First of all came Terray's partial bankruptcy of 1770, in which thousands of government creditors found their incomes slashed. It was the first time the government had defaulted on its obligations for two generations. Over the next few years Terray went on to prolong existing taxes, increase their weight, and begin revising assessments—all in time of peace, when the revenue needs of the government were assumed to be low. And these operations coincided with Maupeou's attack on the parlements, the very intermediary bodies that were supposed to guarantee French political life against despotism. Though dressed up in elevated language, and trumpeted as the preliminary to long-overdue reforms, Maupeou's *coup* originated in court intrigue and was not followed by profound changes. The unscrupulous chancellor and his extortionate colleague were seen as having climbed to power with the support of an unholy alliance of scheming prelates, crypto-Jesuits, and the king's latest mistress, the vacuous Mme du Barry; together they had won over the exhausted *roué* who ruled France, offering him peace and quiet in his old age in return for a free hand to silence every voice of protest or opposition. And yet they succeeded; the parlements were smashed, their ultimate weakness in the face of a determined government vividly demonstrated. The remonstrances of the courts in their own defence proved unavailing. So did an explosion of pamphlets denouncing the changes.[54] The *philosophes* were almost unanimous against the reforms,[55] but even philosophic hostility was unavailing. Despotism had been shown to be a practical possibility in France, and the supposed barriers against it powerless.[56] A major political illusion had been dispelled; and in these circumstances the search began for something better able to restrain the government on a permanent basis.[57] It was no coincidence that the first call for the assembling of the Estates-General to be heard since the last years of Louis XIV came in the remonstrances of the threatened sovereign courts in 1771.

Maupeou's regime disappeared after three and a half years, but its history could not be unwritten. The restoration of the parlements did not solve the problem posed by their previous overthrow. Indeed, their relative quiescence down to the mid-1780s underlined it—they could no longer be relied upon to resist the despotism of the government and its agents. And although Louis XV had died unlamented, and many looked forward to a new age of decency under his successor, hesitations, a distinctly unmajestic bearing, and inability to control the frivolity and indiscretion of the unpopular Austrian queen soon brought Louis XVI into similar contempt. Throughout the 1770s and 1780s the private lives of the royal family and the court were mercilessly pilloried and parodied in innumerable anonymous tracts and scandal sheets. The obscure and disreputable 'diamond necklace' affair of 1785 dragged the queen's reputation down to new depths.[58] The contrast between the contemptible goings-on among the people who ruled France, and the august authority to which they continued to lay claim, was glaring. It made the constant attempts of ministers to extend their power in the king's name seem all the more outrageous and irresponsible. The ever-mounting burden of taxes was odious enough; but seen alongside the frivolous extravagance of a bloated, pleasure-mad court, it seemed doubly intolerable. During the 1770s and 80s, therefore, the informed public had its faith in the way France was governed destroyed, and it was now eager to discuss better ways of doing things, of making government more responsible to national opinion for the way it conducted affairs.

The idea of a national political community had grown over the eighteenth century along with public opinion itself. During the disasters of the Seven Years War the language of an outraged patriotism began to be heard,[59] and by 1770 opponents of Maupeou and the wasteful, corrupt, Jesuitical interests with which he was identified, were calling themselves patriots. They thereby implied that there was a broader national interest to which the chancellor, his allies, and perhaps even the discredited king who sustained them, were enemies. The parlements, by contrast, seemed the spokesmen and the guardians of the national interest, a role they were more than willing to adopt. Through their remonstrances over the century, in fact, they probably did more to imbue the reading public with the idea that the nation was a political entity above the king than any of the handful of writers who discussed such matters.[60] But the writers also played their part, particularly Rousseau, whose *Social Contract* of 1762 argued that the supreme, sovereign authority in a political community was the General Will. The *Social Contract* was, and is, a notoriously difficult and paradoxical book to interpret. We can no more be sure that contemporaries understood it than that we do ourselves. We can be sure, however, that it sold well,[61] and that although Rousseau did not invent the idea of a contract as the basis of political society it was his notion of it that was freshest in people's minds as they discussed the political implications of Maupeou's experiment with despotism.[62] The question Maupeou had brought into relief was that of sovereignty. The political nation had thought itself, and its property, protected from arbitrary power by the balance between the king and the parlements. They constantly clashed, but neither side went beyond certain tacitly recognized limits—until 1771, when these conventions were shattered. The king now renounced the balance, and openly claimed undivided sovereignty for himself. Public discussion increasingly turned to ways of recovering it for the nation at large.

It was generally agreed that the best prospects lay in some form of representative institutions. Between 1771 and 1789, they occupied an increasingly important place in public discussion, and indeed in ministerial experiment. Even before 1771, occasional calls had been heard for representative bodies to be established on a provincial basis. The physiocrat Mirabeau, hoping to lure absentee nobles back to their neglected estates, had called as early as 1750 for the generalization of provincial estates to oversee local government.[63] Throughout the 1760s, the physiocratic journal the *Ephémérides du citoyen* proclaimed the value of local representative institutions. Appalled by the tax increases of the late 1750s, several parlements in *pays d'élections* toyed with the idea of restoring lost estates in order to reassure taxpayers that their interests were being protected; and some echoed these demands during the crisis of 1771. By now, however, the Paris *cour des aides*, led by Malesherbes, and certain provincial parlements like that of Rouen, were calling for national representation in the form of the Estates-General. The parlements, they argued, were merely trustees voicing public concerns in the absence of this ultimate national body. It is true that these arguments failed to launch a sustained campaign for national representation; but they put it on the political agenda, and appeals for the Estates-General continued to be heard sporadically throughout the next fifteen years.

And meanwhile interest in provincial representation grew rapidly. Turgot, a physiocrat in power, commissioned his chief adviser Dupont de Nemours in 1775 to draw up a plan for representative assemblies of property owners at parish, local, and provincial level, each elected by the level below.[64] He fell before taking any practical steps to implement such a plan, but by then the idea of provincial estates had also been discussed in the parlement of Paris,[65] and publicly espoused by the *cour des aides* under Malesherbes. In subsequent years it was also embraced by provincial parlements like those of Grenoble and Bordeaux.[66] Then in 1778 Necker began to introduce *administrations provinciales*—provincial assemblies—whose function was to assist the intendants in their collection of taxes and general administration. Only two of these assemblies were established before Necker fell. A projected third was abandoned along with his other policies. But the original two continued to function down to 1790[67]— the first overt recognition by the monarchy that the governed had any right to be represented in the processes of government.

Public opinion, however, was far from satisfied. The members of Necker's assemblies were either nominated or co-opted by nominees, not elected. Not surprisingly, they came under immediate suspicion of being little more than puppets of the intendants; whereas the estates favoured by various parlements would be beyond ministerial meddling. A quarter of Necker's assembly members were clergy, a quarter nobility, and a half members of the third estate; and although there were precedents for these proportions in bodies like the estates of Languedoc, doubts were expressed as to whether the clergy deserved as much as a quarter of the seats, or whether the nobility did not deserve more. Finally, during the intrigues leading up to the fall of Necker in 1781, his enemies published the text of his original proposal of 1778, which showed that his main aim was to set up a countervailing force to the parlements. Not only did this outrage these bodies and lead that of Paris to obstruct his third proposed assembly, it also sowed the suspicion that his whole intention was to increase governmental power by removing existing obstacles, rather than to give the governed more chance of influencing it. The fact is that, although there was a growing consensus that some representative element ought to be introduced into French government, there was no general agreement about either the form such representation ought to take, its powers, or its procedures. Even the most accessible example of representative government, that of England, much praised as a superb working model since the

Seven Years War, fell into discredit in the 1770s as a result of its failure to deal with the problem of America.[68]

The successful revolt of the thirteen American colonies against British rule added a final element of effervescence to public opinion. No doubt the original sympathy of the French public for the American cause, like the eagerness of their government to give encouragement and help to the rebels, sprang from the desire to be revenged on the British for the humiliations of the Seven Years War. And no doubt this continued to be the French government's primary objective throughout. By the time a formal alliance was struck between France and the new United States, however, French interest in America had passed far beyond the desire to spite the British. People had begun to see in America all their dreams come true—a simpler, healthier, more virtuous society, established in a virgin land of limitless possibilities; a new nation constituting itself from first principles.[69] Much of what Frenchmen thought they saw in America bore little enough resemblance to reality, but the message was no less powerful for that. Enthusiasm for all things American engulfed the country. Benjamin Franklin, the American ambassador and the very embodiment of the new nation's simple virtues, became the most sought-after man in Parisian society. Between 1775 and 1787, the public was deluged with writings on all aspects of America, and showed an insatiable appetite for more. Only the onset of France's own political crisis redirected its attention. By that time the principles and problems of American society and American politics had been thoroughly discussed and absorbed into the way the French public viewed its own affairs. America showed that new starts could be made, that it was possible to renounce a whole old order and set up a new and improved one, that piecemeal reform was not the only road to improvement. It offered the prospect, hitherto merely a matter for theoretical speculation, of a nation establishing itself on the principle that the people were the ultimate sovereign power. It stood as the first example of a people explicitly dedicating itself to the pursuit of political and religious liberty, political equality, and elective, representative government. And not least, perhaps, it filled men anxious to shine and dedicate themselves to great causes with the desire to promote the same aspirations at home. The crisis that unexpectedly occurred in 1787 was to provide them with ample opportunity.

5. Reform and Its Failure 1787–1788

Calonne's plan of reform, approved by Louis XVI after several months of persuasion during the autumn of 1786, had three main elements.[1] First came fiscal and administrative reforms designed to remedy once and for all the structural problems besetting the royal finances. Calonne proposed to recast the tax system by abolishing the *vingtièmes* and substituting for them a permanent, proportional 'territorial subvention' or land tax, to be levied in kind at the moment of harvest. There were to be no exceptions or arrangements for compounding, such as were enjoyed by the clergy or the *pays d'états* under the *vingtièmes*. From this reform Calonne expected an initial increase in revenue of 35 millions; but with the addition of other measures, such as a new stamp duty, more efficient management of the royal domain, and debt redemptions, he expected to increase the crown's revenues far more spectacularly. The land tax was the crucial change, however, and he proposed to assure its success by allowing the landowners who would bear the brunt of it a major role in its administration. There was to be a network of assemblies elected by landowners at parish, district, and provincial level throughout the *pays d'élections*; the *pays d'états* were to keep their existing estates. These provincial assemblies would have an important role in assessing and distributing the weight of taxation and administering public works—but always subject to the supervision and agreement of the intendants, who would remain the prime agents of the central government in the provinces.

Secondly, Calonne believed that a programme of economic stimulation would increase yet further the already improved tax yield to be expected from the administrative reforms. Advised by Turgot's old collaborator, Dupont de Nemours, Calonne proposed to remove a whole range of what physiocrats regarded as impediments to agricultural production. Thus he proposed the abolition of internal customs barriers—

a dream that went far beyond the physiocrats to the time of Colbert himself.[2] Following Turgot, he hoped to abolish the *corvée*, or forced labour for road building, substituting for it an extra tax. There had in fact been several experiments along these lines at provincial level over the previous decade.[3] Finally, he proposed once more to relax governmental controls on the grain trade, going further than either Turgot or his predecessors of the 1760s in allowing free export both internally and externally.

But all these reforms, whether administrative or economic, could not be expected to show instant results, whatever their long-term benefits were expected to be. Calonne had also, therefore, to find some means of surmounting the financial crisis which had brought him to the point of proposing comprehensive reforms in the first place. The immediate problem was to find the money to pay off the short-term debts falling due for redemption between 1787 and 1797. Calonne proposed to do it by raising yet more loans, confident that these would be easy to repay in their turn from increased tax revenues later on as his reforms began to have an effect. The difficulty was to make potential lenders share this confidence, when their evident lack of it had helped to precipitate the crisis in the first place. The comptroller-general's solution was to invest his reform plans with a convincing show of national support, a guarantee that they would go through without obstruction. This ruled out recourse to the most august forum of national opinion, the Estates-General. Calonne regarded 'the representatives of the Nation who remonstrate, who demand, who consent'[4] as far too unpredictable. Instead, he proposed to convoke an Assembly of Notables, royal nominees but 'the principal and most enlightened persons of the Kingdom, to whom the King deigns to communicate his views and whom he invites to apprise him of their reflexions'.[5] They must be 'people of weight, worthy of the public's confidence and such that their approbation would powerfully influence general opinion'.[6] After discussion and approval by such an august body, surely nobody would presume to question the minister's plan of reform further. The normal organs of opposition would thereby be outflanked, and investors given an impressive demonstration of France's determination to restore order to her finances.

It is clear that Calonne did not expect serious opposition to his proposals from the Notables. The proposals constituted after all, as Talleyrand put it, 'More or less the result of all that good minds have been thinking for several years.'[7] Each member of the Notables would be hand-picked, and some of the secondary proposals were carefully framed to conciliate interest groups who might feel affronted by the major ones. Noblemen, for instance, would now be exempted for the first time from the *capitation*, and there was no question of subjecting them to the *taille*, the basic direct tax, or any levy designed to compound for the *corvée*, from which they were exempt. The only direction from which Calonne feared serious obstruction was from his own ministerial colleagues; but with the support of the powerful foreign minister, Vergennes, and the king himself, he felt strong enough to proceed. On 29 December 1786, the convocation of the Notables was announced. A list of 144 members was drawn up, comprising 7 princes, 14 bishops, 36 titled noblemen, 12 members of the Council of State and intendants, 38 magistrates, 12 representatives of the *pays d'états*, and 25 mayors and civic dignitaries. After several delays, the Assembly of Notables eventually opened at Versailles on 22 February 1787.

It became clear almost from the start that Calonne had badly miscalculated his ability to control the Assembly. He had underestimated three crucial factors. First came the determination of the clergy and the *pays d'états* to preserve their privileged position in the state. The representatives of these two elements were few in number, but their determination and their skill and experience in manipulating their own assemblies made them formidable organizers of opposition. The clergy in particular was threatened by his plans, and not only by the intention to abolish its exemption from direct taxes. Equally obnoxious was a concomitant proposal to make the clergy redeem its corporate debt by selling off various properties and feudal rights so that it should no longer be able to justify its tax-exemption by claiming that it helped the state by borrowing. The prelates declared themselves unable to assent to anything so radical without the approval of the Assembly of the Clergy, which was not due to meet until later in the year. They were also the main source of most of the objections raised by the Notables to the form and powers of the provincial assemblies, and the justice and practicability of the land tax. Even so, the opposition of the clergy to every aspect of the plan which involved them would have been far less serious were it not for the persistence of opposition to Calonne personally from other quarters. Not only were there personal enemies, like the first president of the parlement of Paris, within the Notables. Ministerial rivals, such as the keeper of the seals, Miromesnil, viewed the Assembly more as a means of bringing Calonne down than of confronting the serious problems that beset the state. Nor were such plottings now restrained by the prestigious Vergennes, who had died just before the Assembly met. And meanwhile, certain prominent members of the Assembly saw it as an opportunity to advance their own ambitions by discrediting Calonne and perhaps replacing him—the most noteworthy of these being the worldly but able Archbishop of Toulouse, Loménie de Brienne. There was also a vocal and well-organized Neckerite party, anxious to see the former minister restored to power and to defend their hero against the charge, implicit in much of Calonne's analysis of the state's problems, that he had been largely responsible for them. All these elements were only too eager to seize upon the criticisms of the ministerial proposals advanced by the clergy, in order to undermine the comptroller-general. And they were helped by Calonne's difficulties in establishing his own credibility. He claimed that the state was facing an unprecedented crisis in its finances, and yet at first he refused to divulge detailed accounts to substantiate this. He had always enjoyed a public reputation of being too clever by half, and it is not, therefore, surprising to find many of the Notables somewhat sceptical about the need for such radical changes as were now being canvassed. These doubts, assiduously fostered by those who wished to bring Calonne down, snowballed during the first week of the Assembly. By the beginning of the second week Calonne was forced to recognize that he would get nowhere unless he gave more details of the state's financial condition. But when he reluctantly declared that he expected a deficit of 112 millions on the year, the Notables were still not satisfied. They now demanded free access to the complete accounts, and meanwhile asked whether the real cause of the deficit was not the lavish and ostentatious expenditure incurred by the minister since 1783. When Calonne implied that the true blame lay with his predecessors, the Neckerites pointed out that their man had declared a surplus in his *Compte rendu* of 1781. It was difficult, and it still is difficult, to know whom to believe.

What is certain is that Calonne's whole strategy misfired. He had expected the Notables to accept that there was a crisis, accept his diagnosis of its causes, and endorse his proposals for dealing with it. They had done none of these things. This is not to say that they had set their face against all innovations. It seems clear that, if it could once be proved that there was a genuine crisis, the Notables were prepared to think radically about how to overcome it. Almost unanimously they agreed that taxation should henceforth fall equally on everybody within the state; the clergy and the *pays d'états* representatives merely observed that to be equal in weight taxes did not need to be uniform in their collection. The representatives of the titled nobility were insulted at the proposal to exempt them from the *capitation*, seeing it as a low bribe to induce them to accept other measures whose importance they were quite public-spirited enough to appreciate. Everybody agreed on the desirability of *some* form of provincial assemblies, of the commutation of the *corvées*, the emancipation of the grain trade, and a number of other measures. Most welcomed the abolition of internal customs barriers. There is no reason to impugn the sincerity of these professions; and if some at least of the practical criticisms levelled at the land tax and the proposed form and powers of the provincial assemblies were primarily designed to embarrass Calonne and cripple his major proposals, it would be naïve to conclude that all those convinced by the criticisms were as Machiavellian as their clerical originators. The difficulties were genuine; and so were the Notables' fears about Calonne's ways of proceeding. Though making a display of consulting with the cream of the nation, he seemed quite unwilling to accept any of their criticisms. They did not forget that the minister was a former intendant, and suspected that his true preference was for the despotic ways of those officials. The proposed subordination of the provincial assemblies to an intendant's veto in all matters of substance suggested that he had little enough faith in true consultation. Yet the measures now proposed were so momentous that only a free show of national approval would be enough to legitimize them. Various magistrates gave notice that they had no power to commit the parlements in which they sat to register the reforms uncritically; that, they said, would be to betray the trust placed in them by the public. They had, in fact, declared the procurator-general of the parlement of Aix, no authority to approve the land tax. An innovation of this importance needed nothing less than the consent of the Estates-General.[8]

This lead was not immediately followed. But when on 12 March Calonne blandly declared that the king was satisfied to see that the Notables agreed on all the basic principles put before them, they exploded with indignation. Nothing could have been further from the truth, at least in the case of the land tax, whose very necessity more and more of them were coming to doubt. There seemed little to be gained from talking to a minister who only wanted to hear his own views confirmed; and almost total non-co-operation now became the order of the day. Calonne, sensing that his policies were losing momentum and the king's confidence beginning to waver, now made a bold attempt to appeal over the Notables' heads to public opinion. There had been nothing secret about the convocation of the Assembly, but until now neither the nature of the financial crisis nor the minister's proposals for resolving it had been made public. The public was agog to know what was happening, but had to content itself with gossip about personalities or unauthenticated leaks. Calonne now abandoned this secrecy, published the text of his proposals on the land tax and the provincial assemblies, and accompanied it with an inflammatory *Avertissement* which denounced the Notables' criticisms as the self-interested obstructionism of privileged groups indifferent to the wider national interest. It was distributed free, in massive quantities, and parish priests were urged to read it from the pulpit. It was an obvious attempt to cow the Notables into acquiescence by mobilizing public pressure, the most ambitious attempt to woo the French public since Necker's *Compte rendu*.

But it failed dismally.[9] There was no response to the inflammatory incitements against the privileged—it was to be another eighteen months before the war on privilege really broke out. Meanwhile the public saw in Calonne's plan only the prospect of higher taxes. If there was a crisis, they were inclined to think like the Notables that the minister himself must have been responsible, since Necker had shown that all had been well in 1781. Necker himself rushed into print with a pamphlet defending his record, while other writers implied that Calonne was deeply implicated in shady dealings on the stock market. The Notables, for their part, were outraged at his attempt to change the rules of the game half way through it; and his evident failure only strengthened their determination not to co-operate with him in any way. Calonne, in short, instead of refurbishing his credit had destroyed its last shreds, and now all his enemies combined to deliver a *coup de grâce*. From ministers, from the royal family, and, via the queen, from prominent Notables like Brienne, the king was bombarded with denunciations of his comptroller-general. Having been convinced himself by Calonne's proposals, Louis was reluctant to abandon either them or their author; but he now realized that he must choose one or the other. On 8 April, in the hope that some other minister might yet carry the reforms through the Notables and into effect, Louis dismissed Calonne. As if to demonstrate that he was still determined to brook no opposition to the reforms themselves, Louis had also dismissed their most insidious critic, Miromesnil, earlier in the day.

Yet this resolved no basic problems. The Notables were still out of control, and the financial crisis which had led to their convocation in the first place was getting worse as no steps were taken to resolve it. The appointment of a collaborator of Calonne, Fourqueux, to succeed him, did nothing to improve matters, and the demand began to grow among the Notables for the convocation of the Estates-General.

Only when the king began to take the advice of Loménie de Brienne, and promise modifications in Calonne's programme, did they appear more co-operative. Brienne's personal ambition was blatant, and Louis was reluctant to gratify it. But the results of taking his advice were so striking that he was left with little choice, especially in the light of steadily declining royal credit. To avoid bankruptcy, therefore, and to regain control of the Notables, on 30 April Louis appointed Brienne first minister. Another Notable and known advocate of legal reform, President de Lamoignon of the parlement of Paris, had already succeeded Miromesnil as head of the judiciary.

This was the ministry that was to precipitate the collapse of the Old Regime. And yet it began promisingly enough. Advocates of legal reform were ecstatic at the fall of Miromesnil and the appointment of Lamoignon. The news of Brienne's elevation to the ministry produced a dramatic rise in royal stocks on the money markets. On Brienne's advice, the fullest possible statement of the royal accounts was now released to the Notables, and they were able to satisfy themselves at long last that the financial problem really existed. The king now offered, in response to the Notables' previous criticisms, to tailor the land tax according to each year's needs, to give it a fixed duration, and to levy it in cash; and to modify the composition of the provincial assemblies in order to guarantee a minimum number of seats to the nobility and clergy. He also offered to make further economies and to spread the burden of debt redemption over a longer period. Brienne hoped that these concessions would swing the Assembly back behind the government and so overawe any potential future resistance to its plans. But the Notables had gone beyond merely responding to government initiatives. They now had ideas of their own. To prevent any recurrence of the financial disorder that had brought them together, for instance, they proposed a permanent commission of auditors to supervise the government's conduct of its financial affairs. When this was refused (by the king himself), they began to call once more for the Estates-General. All the government wanted, more and more of them began to say, was a free hand to tax at will, and this they refused to connive at. 'We did not think', declared one subcommittee typically enough on 21 May, 'that an Assembly of Notables, without power and without a mission, which had not been deputed by the provinces and which had nothing in common with the Estates-General, could vote a tax.'[10] If it did so, declared a member of another subcommittee, the Assembly would 'compromise itself in front of the Nation.'[11] Clearly matters had now reached a dangerous stalemate. The Notables were close to declaring that there could be no new taxes in the absence of consent from the Estates-General. Brienne, the king, and the other ministers, already chastened by the experience of dealing with a hand-picked Assembly, were quite determined not to call an elective one. Yet without some new taxation the finances could never be put back in order. The second attempt to win endorsement of the government's reform proposals had failed, and by the end of May Brienne recognized the fact and decided to cut his losses. On 25 May, having given its unequivocal consent to hardly any of the government's proposals, the Assembly was dissolved, and Brienne prepared to promulgate his programme regardless.

The intransigence of the Notables should not be seen as the unwillingness of the 'privileged orders' to contemplate reform.[12] To do so is to think in the terms of 1789 rather than 1787. By May at least, when the government belatedly agreed to give them access to all its accounts, most of the Notables recognized that there was a crisis and that far-reaching changes were necessary in order to meet it. But they concluded (and who are we to say that they were wrong?) that the disorder in the finances was the result of incompetent government; and quite reasonably they asked for independent safeguards against further incompetence. They were offered none, and this is why they fell back on the time-honoured idea of the Estates-General and refused to transact other business. And meanwhile, in consequence of the fatal miscalculation that the Notables would be easy to control, the government had irrevocably compromised itself. In calling the Assembly at all, it had implicitly admitted that it needed some show of national consent to sustain its credit. Its failure to win the Notables' support showed the world that the

French king's government could not command the confidence of its most eminent subjects; while the revelation of the disorder in the finances suggested that this lack of confidence was well merited. By attempting to stampede the Notables with his appeal to public opinion in March, Calonne had turned the crisis into a matter for general discussion rather than a private exchange of views between the king and chosen advisers; and the Notables were reinforced in their intransigence by a public opinion as sceptical of the government's intentions and competence as they were themselves. The Assembly had also, in declaring its own inability to sanction anything, given renewed impetus to the idea of the Estates-General, which had certainly been in the back of people's minds since 1771, but had never before been a constant rallying cry. And finally, the experience of the Notables compromised Brienne. It had indeed proved his ladder to power, but in leading the opposition to Calonne's plans, only to adopt them in a modified form once in office, he discredited himself in the eyes of the public almost before he began to exercise power. In all these ways, by the summer of 1787 the political atmosphere of France had changed beyond recognition. The pre-revolutionary struggle had begun.

It was unfortunate for the Brienne-Lamoignon ministry that it had to work in such a highly charged atmosphere, for its programme promised to make it perhaps the greatest reforming government of the century.[13] Lamoignon embarked on a major project for reforming and recodifying the laws, and began by abolishing the last vestiges of judicial torture. He also launched studies on reforming the educational system. A start was made on introducing religious toleration by granting a civil status to Protestants. Measures were taken for streamlining the army so that it cost less but performed more efficiently and more professionally. In the central government, the Notables' idea of a panel of auditors was taken up in the form of establishing a new central Council of Finances to oversee the activities of the comptroller-general; and the process of eliminating private enterprise in the administration of the finances, abandoned since the fall of Necker, was resumed with the creation of a central, bureaucratically run treasury with a clear annual budget.[14] Unfortunately, most of these reforms could not be divorced from those which had already been savaged by the Notables, but with which Brienne was nonetheless determined to push ahead. To implement them legally, without recourse to the Estates-General, there was no alternative to having them registered by the parlements.

Brienne had no illusions about what would happen if the government failed to win the Notables' support. 'If the Assembly', he wrote on 16 April, 'disperses without having assured a balance between receipts and expenses and one is obliged, after its dispersal, to have recourse to taxation, it is to be feared that great resistance will be encountered. This Assembly was formed because it was judged that its opinion would vanquish all the obstacles that could be foreseen.'[15] In the event, its hostility to what was proposed had the opposite effect. The Notables had implied that only the Estates-General could legitimately consent to new taxes, and they had sowed the same idea among the public. The parlements, whatever their own views, could hardly be expected to adopt a less extreme attitude; and when in the course of June 1787 the government began to send its reforming edicts for registration, its worst fears were realized. Even so the parlements were not totally intransigent. They raised no trouble over registering the emancipation of the grain trade or the commutation of the *corvée*, and little over civil status for Protestants. Most of them even failed to demur at registering the edict instituting provincial assemblies, although that of Bordeaux, long on record as an advocate of revived provincial estates rather than administrative assemblies, refused point blank to register until the government clarified the form and powers of the proposed assemblies in the south-west—which it refused to do.[16] The sticking point for most of the parlements, however, came with two edicts on taxation—the one replacing the *vingtièmes* with a modified land tax, the other introducing a higher stamped paper duty. And here, for the first time since 1776, the lead in opposition was taken by the parlement of Paris, reinforced by an influential body of former Notables who sat by right in the parlement as peers of the realm, and urged on by the public opinion of the capital, which was practically unanimous in its hostility to the ministry. 'The parlement of Paris', observed Malesherbes, who had joined the ministry at Brienne's invitation, 'is, at this moment, but the echo of the public of Paris, and . . . the public of Paris is that of the entire Nation. It is the parlement which speaks, because it is the only body that has the right to speak; but let there be no illusion that if any assembly of citizens had this right, it would not make the same use of it. So we are dealing with the entire Nation; it is to the Nation that the king responds when he responds to the Parlement.'[17]

When the stamp duty edict was presented on 2 July, the parlement refused by an overwhelming vote to consider it until the government presented a full set of accounts justifying its necessity. A fortnight later the land tax met a similar refusal, and a few days before that the magistrates issued a remonstrance in which they called for the convocation of the Estates-General. Only the Estates, the magistrates now explicitly declared, could give a legitimate consent to new taxes. The seriousness of the parlement's opposition was not yet clear. The British ambassador wondered 'whether it is only to demonstrate to the World for the sake of Parliamentary dignity that they do not in everything implicitly obey the King's commands'.[18] But unofficial attempts to reach a compromise failed, and on 6 August the two tax edicts were forcibly registered at a *lit de justice*. In order to soften the blow, new economies in the royal household were announced the same week, but the parlement was unimpressed. It declared the proceedings of the *lit de justice* null and illegal, and began criminal proceedings against Calonne. Jubilant daily demonstrations left no doubt of the popular support the parlement was now receiving, as its genuine opposition to the ministerial plans became clear. The government became alarmed. An international crisis was coming to a head in the Dutch republic, where 'patriots' opposed to the Prince of Orange and hitherto supported by France, seemed on the point of provoking a Prussian invasion. The British were making it clear that any French intervention on the patriots' behalf would mean war—and war, of course, would mean enormous new expenditure. In an attempt to resolve matters quickly on 15 August, the king exiled the parlement in a body to Troyes until it should prove more compliant. Four days later he ordered the closure of all political clubs and discussion societies in Paris, bodies which had mushroomed since the spring, and which ministers suspected of co-ordinating many of the pro-parlementaire demonstrations. On 26 August, Brienne's ministerial strength was shown when he took the title of principal minister and forced the resignation of two hostile colleagues. The ardour of the magistrates certainly cooled in Troyes, and Paris, too, calmed down after the closure of the clubs and a determined attempt to stamp out unauthorized publications. But the provinces were now involved because the parlements of the regional capitals in their turn refused to register the new edicts. That of Bordeaux preceded Paris into exile for resisting the provincial assemblies. All this kept potential waverers at Troyes firm. The parlement refused to publish the forcibly registered edicts and ordered its subordinate courts to do likewise. In these circumstances the new laws could not take effect, and with no promise of increased tax yields, there could be no question of war. When on 13 September the Prussians finally crossed the Dutch frontier, the French stood aside and sent their 'patriot' allies no help. Within a month the whole of the Dutch republic was occupied. France, the arbiter of Europe in the time of Vergennes, had been reduced by her internal crisis to an international cipher.

Brienne was opposed to intervention in the Netherlands in any case, but he could hardly consider the financial inability to intervene satisfactory. If France's power were not to suffer permanent damage, the constitutional stalemate had to be broken. Accordingly, in the course of September, he elaborated a new plan which he hoped the parlement would accept. Abandoning the stamp and land taxes, he now simply proposed to extend the existing *vingtièmes* until 1792, but with provisions for accurate assessment and no exceptions or special cases. Along with economies in expenditure, he calculated that in five years this proposal could restore the finances, provided that further loans could be raised in the interim to cope with repayments falling due. So confident was he in this plan that he also offered a major concession. By 1792, he declared, the Estates-General would be convoked to celebrate financial recovery. This swung the parlement over; honour was satisfied; no new taxes (merely old ones prolonged) and a promise of a meeting of the Estates. The *vingtième* prolongation was quickly registered, and in return the government revoked the exile to Troyes. The settlement would be confirmed at the beginning of the new judicial year by the registration of the new loans.

'Patriotic' opinion was suspicious: 1792 was a long way off, and meanwhile further taxation had been sanctioned. The timorous magistrates seemed, after all, to have sold out to despotism. The provincial parlements, too, openly expressed their disappointment. The anti-ministerial coalition appeared to be falling apart. What saved it from dissolution was the unforeseen course of the sitting of 19 November, when the new loans were to be registered. This was planned as a 'Royal Session', but not a *lit de justice*. Members of the parlement were to opine freely, in the monarch's presence, about the loans. In the course of a day-long session, all shades of opinion found expression, but on a vote the loans would probably have been registered.[19] But no vote was allowed. Instead the king, as in a *lit de justice*, simply ordered registration. The duc d'Orléans, expressing the instantaneous reaction of the stunned company, shouted out that this was illegal. Louis, equally startled, snapped back, 'I don't care . . . that's up to you . . . yes . . . it is legal because I wish it.'[20]

This was despotism indeed, and it seemed to confirm all the worst suspicions harboured by the public ever since the convocation of the Notables. The government constantly spoke the language of conciliation and consultation, but was nevertheless determined to get its way using the full weight of its authority. The lesson was only underlined the next day, when Orléans was exiled by *lettre de cachet* and two leading magistrates were imprisoned. A further demonstration came from the provincial assemblies, which now were meeting for the first time and dealing with the *vingtièmes* prolonged in September. Their members were government-nominated on this first occasion, and Brienne persuaded a number of them to agree to compound for their provinces' contribution so that they would have to raise an agreed sum rather than whatever two-twentieths brought in. This was to reintroduce the repartitional principle of the abandoned land-tax.[21] Provinces whose assemblies refused to agree were subjected to punitive reassessments, and they did not hesitate to complain of the fact to the parlement. All these things outraged the magistrates and made them determined to have no further truck with the ministry. Between November 1787 and May 1788 the air was full of acrimonious exchanges. A series of remonstrances denounced *lettres de cachet* and arbitrary imprisonment, but the king remained unmoved. The new taxation procedures were also denounced, and from the provinces the rest of the parlements swelled the chorus of protest. Only in a few areas were the new *vingtièmes* registered, and in several others the provincial assemblies were prevented from meeting. The only legitimate form of provincial representation, most of the courts declared, was that of estates, and where those were defunct local parlements called for their re-establishment. In these months, in fact, the parlements came closer to acting as one body than they had ever done before. Together they reduced the ministerial plans, whose very essence was uniformity of application, to chaos.

The time for conciliation was now obviously gone. The Royal Session had destroyed any hope of that on both sides. In the course of the spring, accordingly, ministers began to elaborate a final solution to the constitutional deadlock, a Maupeou-like reform that would curb the parlements' powers of obstruction once and for all. Long before it happened there were ominous rumours; and these only encouraged the magistrates to adopt ever-more extreme positions, denouncing the blow that was about to fall and enunciating the principles of their opposition for posterity. This was the atmosphere that produced the most extreme constitutional statement of the century from the parlement of Paris on 3 May 1788. Throughout the century in their remonstrances they had invoked 'fundamental laws' of the kingdom which no king had the right to transgress. Now for the first time they presumed to list them. Among principles which nobody would care to dispute (such as the salic law of succession), the parlement also now placed maxims scarcely spoken of until

1787, such as that taxation must have the free consent of the Estates-General meeting regularly, and that all arbitrary arrests and imprisonments were illegal. The magistrates engaged themselves to refuse all co-operation with any new order that might be introduced. Yet the ministers appear to have been confident that their *coup* would succeed. Maupeou, after all, despite an immense furore, had succeeded in bridling the parlements before; and since the new loan forcibly registered at the Royal Session of November 1787 had been quickly subscribed, they expected none of the financial worries that had forced them to compromise the previous September. On 6 May, they arrested the two most fervent of their parlementaire opponents, d'Eprémesnil and Goislard, in open parlement, despite the court's attempts to protect them. Then on 8 May, at a *lit de justice* timed to coincide with military sessions at all the provincial courts, the keeper of the seals, Lamoignon, unveiled his reforms and had them forcibly registered on the spot.[22]

Henceforward, all general laws were to be registered only by a Plenary Court consisting of senior magistrates, great nobles and officers of state, and a collection of eminent persons not unlike the membership of the Assembly of Notables. This Plenary Court alone would have the right to remonstrate against general laws. The role of the parlements was therefore restricted to registering laws of purely local character; their traditional political role was annihilated. And their judicial role, too, was now to be changed beyond recognition. Lower courts, notorious for their resentment at the tutelage in which the parlements kept them, were now to be selectively elevated to the status of *grands bailliages*, enjoying competence over far more important cases than they had handled before, at the expense of the parlements who had hitherto monopolized them. This was the essence of the reform; but in order to make it more attractive, Lamoignon also introduced at the same session laws abolishing various special courts which needlessly complicated the machinery of justice, and a hastily framed Criminal Ordinance, the first fruit of his plans for reform of the criminal law. These measures were now simultaneously registered by force in all the country's sovereign courts, and so were all the various edicts rejected until now by difficult parlements. The magistrates were given notice that, in view of the diminution of business that they could now expect, the number of their offices would shortly be drastically cut. In order to preclude any protests, the courts were all then put into vacation until further notice.

The whole country now exploded with indignation. All the parlements except Paris ignored the order to go into vacation, and passed resolutions declaring the forced registrations null and illegal. Several continued to sit and hear cases. When the government imposed obedience by *lettres de cachet*, supported by the threat of troops, the departure of the magistrates from their various palaces of justice took place amid huge anti-ministerial demonstrations. At Rennes and Grenoble troops had to be called in, and in the latter they were bombarded with tiles from the roof-tops. Meanwhile the public was inundated with copies of remonstrances, resolutions, and protestations issued by the courts, and these were followed by a wave of pamphlets dissecting the reforms —the vast majority, with the exception of those commissioned or subsidized by the ministry, deeply hostile. The attempt to win over opinion by decorating the changes with the promise of legal reforms was even less successful than it had been under Maupeou. Enlightened ends could never justify despotic means. The time had come for the Estates-General to pronounce, not just on new taxes, but on a whole range of reforms which, however desirable, could only be made legitimate by its consent. The social contract had been broken, the general will defied. It was time to give France a new constitution. Such themes occurred again and again in the pamphlet literature,[23] while defences of the ministry were scorned. Meanwhile, there was a general refusal to cooperate in making the reforms work. Most of those designated as members of the Plenary Court refused to serve, and only in a few inferior courts, dazzled by promotion to the status of *grand bailliage*, did the government find willing collaborators.[24] The Assembly of the Clergy denounced the reforms, called for its privileges to be respected, and voted a *don gratuit* less than a quarter of what the government had requested. And all over the country noblemen met together in unauthorized assemblies to organize petitions and discuss other common action in favour of the stricken parlements and the convocation of estates, both provincial and general.[25]

All this action undoubtedly created chaos, and threatened more. But whether it would have deflected the government from its course in the long run, any more than the outcry of 1771 had done, is open to doubt. The opposition was noisy, but only in Dauphiné did it defy authority actively by resurrecting the defunct provincial estates on its own initiative.[26] The armed forces, the government's ultimate weapon, remained basically loyal and obedient, despite some discontent among officers.[27] There were enough signs of division and uncertainty among the forces of the opposition to make it likely that, had the ministry been able to persist with its policies for more than a few months, the clamour would have died down just as it had in the early 1770s. After the first outcry, Brienne even took the offensive with an attempt to split the opposition further. On 5 July he announced that the government was prepared to consider all suggestions about the form the Estates-General should take, when they met. He invited interested parties to submit their views before the spring of 1789. Not only would this buy him time; it was also likely, indeed calculated, to sow confusion and division in the 'patriotic' ranks which would leave the ministry free to organize the Estates as it liked, if indeed it did not provide an excuse to avoid convoking them at all. The whole history of the eighteenth century had shown that a united and determined government had little to fear from any opposition. In July 1788, Brienne looked set to emphasize this lesson once again.

What prevented this was the sudden collapse of the government's finances. The financial problem had precipitated the constitutional crisis in the first place, and now it was to resolve it, too. The government was safe as long as it could borrow money, and its credit was still good enough in November 1787 for even a forcibly registered loan to succeed. But the day-to-day finances of the government did not depend just on public loans. They also depended on *anticipations*, short-term credits from bankers and financiers secured on the tax revenues of subsequent years. Brienne hoped in the long run to eliminate the need for *anticipations*, but the budget for 1788 was still burdened with 240 millions due from previous years. This had to be made up by anticipating the same amount from future years, but this time it proved impossible. The government's usual creditors refused to lend, no doubt because their liquidity was restricted by a burgeoning economic crisis,[28] perhaps because they had no interest in supporting a minister who was planning to dispense with their services, and certainly because the *coup d'ètat* of May had at last shattered the government's credit. The despot who ruled France could no longer be trusted with good money. Yet without *anticipations* the government could no longer cover its daily expenses. At the beginning of August, the comptroller-general told Brienne that the treasury was empty. On the 8th, in a last desperate attempt to reanimate credit, Brienne fixed a precise date for the meeting of the Estates—1 May 1789—thus giving up at a blow his whole delaying policy. But even this had no effect. Confidence was not restored, and a week later, on 16 August 1788, Brienne was forced to suspend payments from the royal treasury. Creditors were now compelled to take interest-bearing notes—a sort of forced loan.[29]

Recently it has been argued that this was not really a bankruptcy, but rather the logical culmination of two decades of experiment with eliminating the role of private business in public finance.[30] But contemporaries certainly took it for a bankruptcy, and that is what mattered at the time. There was a run on the *caisse d'escompte*, and no sign that the money market would do anything further to help the crown so long as Brienne remained in power. The ministry was doomed, and with it all its plans of reform. Brienne's final move was to engineer, despite Louis XVI's personal hostility, the return to power of the one man whose reputation for financial wizardry had remained intact throughout two years of controversy—Necker. Then he resigned, and Lamoignon followed him a few weeks later. Like Maupeou and Terray before them, their fall brought the abandonment of their reforms, too. But this time it was more than the end of a policy. The Estates-General were now to meet in 1789, and Necker made it clear that, apart from sustaining credit, he proposed to do nothing until they did so. Having run out of money, the old monarchy and its servants had also run out of ideas. Here, in the last weeks of August 1788, the Ancien Régime, as a political system, collapsed.

B. The Struggle for Power

In August 1788, the old monarchy collapsed. It was not overthrown by the opposition to its policies, much less by revolutionaries dedicated to its destruction. It fell because of its own inner contradictions.

Its disappearance left a vacuum of power. Throughout the winter of 1788/9, Necker ran a caretaker government which saw its only business as supervising an orderly progress towards the convocation of the Estates-General. It would be the task of this body, not of the king and his ministers, to formulate solutions to the state's problems. This was as much as to announce that power in France had passed from the king's hands. The question then arose, naturally, of who was to inherit it: who was to control the Estates-General, the body that was to regenerate the nation? This issue dominated French politics from September 1788 until July 1789. These months saw a struggle for power between groups who felt entitled to a say in how France should be governed from then on. Only when the struggle was over did men turn their attention to what should be done rather than who should do it. And by then the struggle for power itself had produced a whole series of aspirations, priorities, and issues to which little thought had been given before 1788. The French Revolution was the process by which these matters worked themselves out.

6. The Nobility

Like most countries of eighteenth-century Europe, France was dominated by nobles. All the king's ministers were noble—apart from the extraordinary case of Necker. The royal court was exclusively peopled by nobles. All the intendants were noble, as were nearly all the bishops and other high ecclesiastical dignitaries. Almost all army officers, and a substantial proportion of naval ones, were noble. So were all members of the sovereign courts, by definition.[1] Nobles also possessed the lion's share of the country's wealth. They owned between a quarter and a third of the land and pocketed about a quarter of the total agricultural revenue.[2] Through their position in the church they also enjoyed the usufruct of between 15 and 25 per cent of that institution's revenues.[3] They exercised seignorial rights over most of the land they did not own. Most of the immense capital invested in venal offices was noble capital. Most heavy industry was financed by noblemen,[4] and even the world of banking and high finance was full of noble men of business.[5] The only important sectors of economic life that they did not dominate were those of light industry and trade—and their participation was not unknown even there. Indeed, it has been argued that even in the sectors they did not dominate, nobles were the boldest and most enterprising participants, pioneering new techniques and opening up new markets.[6] And finally, nobles dominated France's cultural life. They were the greatest patrons of the arts, the leaders of fashion, and the mainstays of the intellectual world. They filled the academies, national and provincial. The aristocratic ladies of the *salons* were the midwives of the Enlightenment, ventilating the latest ideas, introducing and protecting new writers. A remarkable number of the writers themselves were noble, too: one need only think of Montesquieu, Condorcet, d'Holbach, or Jaucourt. Not all nobles, of course, were men of power, wealth, or influence. But the vast majority of such men were nobles.

We cannot yet be sure how many nobles there were in 1789. Recent estimates vary between 110,000 to 120,000 and 350,000. Either way they constituted a very small proportion of the population—between .5 and 1.5 per cent. Yet in addition to its wealth and power, this tiny fragment of the French nation enjoyed important privileges. It was far from the only privileged group in France. Privilege was universal; and the French king had few subjects who, in virtue of the province or town where they lived, or some body to which they belonged, did not enjoy the right to some sort of special treatment.[7] But because they were so conspicuous, the privileges of the nobility attracted far more notice than those of any other body, except perhaps the church. Nobles enjoyed the right to precedence on public occasions, the right to wear a sword in public, and the right to display a special coat of arms. These were their 'honorific privileges'. More important were the 'useful privileges'. These included the right to trial in special courts, and exemption from billeting, from militia service, from the *gabelle* or salt monopoly, and from the *corvée*. Above all, they included exemption from the *taille*, the main direct tax, where this was levied on persons rather than land—perhaps three-quarters of the country. Until 1695, this had meant that nobles had paid no direct taxes at all. They justified this situation, if called upon to do so, by the medieval maxim that society consisted of those who worked, those who fought, and those who prayed. The first contributed taxes to the common welfare; the other two contributed their services, and the fighters were the nobility. This argument still occasionally surfaced in the eighteenth century,[8] but by subjecting them to the *capitation* from 1695, and the *dixième* from 1710, Louis XIV had destroyed the nobles' tax immunity. Nor, unlike the clergy, were the nobility exempted from the *vingtième* introduced in 1749. Their privileges were not, therefore, threatened by Calonne's proposal to replace the *vingtièmes* with a land tax in 1787. The real threat, equally serious for all landowners regardless of status, came from the proposal to levy the new tax in kind at the moment of harvest; for most of the *vingtième* assessments had not been substantially revised since the mid-1750s, whereas landed revenues had steadily risen. Nobles, however, because of their prestige and contact with local authorities, were far better placed than most of their fellow citizens to evade the full weight of taxes, and many certainly paid far less than they ought to have done.[9] This would have been far more difficult with a tax raised in kind.

Along with privileges went obligations, which most nobles readily acknowledged. They were forbidden by law to engage in retail trade or any manual work, on pain of *dérogeance*, or loss of status. By the eighteenth century this law was so full of loopholes, many of them state inspired, that no noble wishing to trade need seriously fear loss of status for doing so.[10] Far stronger than the law, however, was the prejudice shared by most nobles against trade and manual work. For a noble to stoop to such things was thought dishonourable. The scale of noble involvement in industry and business by the late eighteenth century shows that even this prejudice was on the wane; but the process was slow, and most new recruits to the nobility remained only too eager to abandon their commercial activities in order to speed their assimilation into the social élite. The aim was to 'live nobly', that is to say, without working, sustained by unearned income, preferably from landed estates. Honour dictated more than genteel idleness, however. Nobles also expected to serve the king in order to earn their privileges. The origins of the nobility were believed to be military, and it still regarded itself as the natural supplier of officers for the army. Most officers had certainly always been recruited from its ranks, and over the eighteenth century this tendency grew more pronounced. By the 1780s, it was government policy to restrict recruitment of officers almost exclusively to nobles of established lineage, convinced as the government was that this would produce more dedicated, professional officers[11] —as it certainly had in contemporary Prussia.

Scarcely less prestigious than a military career by the 1780s, however, was one in the judiciary or the administration. Office in the sovereign courts had always conferred ennoblement, but until well into the seventeenth century the 'nobility of the robe' was not really considered the equal of that of the 'sword'. In the Estates-General of 1614, magistrates sat with the third estate rather than the nobility. A century later matters had changed. Nobody now doubted the noble credentials of the members of the parlements,[12] and by the eve of the Revolution only a small minority of them had been ennobled by the offices they held. Most enjoyed nobility beforehand, and several courts refused to recruit non-nobles at all. It would be wrong, therefore, to follow revolutionary propagandists in regarding the nobility as a flock of idle parasites. The high society of Paris and Versailles, and indeed of the great provincial capitals, was full of them, of course. Louis XIV had taken special pains to concentrate his greater nobles around his person at Versailles in glittering impotence, supported by lavish sinecures and pensions from the public purse. But many never reconciled themselves to this role, and over the eighteenth century courtiers gradually worked themselves back into positions of real power and responsibility. In any case, before the late 1780s, those who denounced the spendthrift uselessness of the court most persistently were fellow nobles like the physiocrat Mirabeau the elder, convinced that such displays were unworthy of the Second Order of the state. Few doubted, meanwhile, that without the services of noblemen in administration, the law, and the armed forces, the king's government could not be carried on.

The essence of nobility was that it was hereditary, passed on from fathers to their children in the blood. Some purists accordingly believed that it could never be acquired. For them, what vulgarly passed for the nobility was made up of a tiny handful of true nobles whose lineage was lost in the mists of time, and a vast majority of *anoblis* who could only ape, never join them. But this was never a majority view, for it implicitly denied the king one of his highest prerogatives—that of conferring nobility on his subjects. Throughout early modern times, in fact, the king was lavish with this prerogative, creating huge numbers of offices which automatically ennobled their holders or their descendants. In the eighteenth century, there were perhaps 3,700 ennobling offices, and although a third or more of these were probably held at any given time by men who already enjoyed nobility, the rest were bought by socially ambitious bourgeois intending to ennoble themselves and their families. Nobles liked to think of themselves as an exclusive race apart, and it is clear from their very denunciations of the nobility's apartness that many later revolutionaries believed they were. But in reality the eighteenth-century nobility was an open élite, accessible to anybody who could afford the price of an ennobling office. Attempts are now being made to calculate how many people were ennobled over the century. As in the case of the total number of nobles, disagreements remain. One estimate is 6,500, another nearer 10,000.[13] Multiplied by five for the families who inherited noble status from their newly ennobled heads, this gives a minimum total of 32,500 or a maximum of 50,000 new nobles during the eighteenth century—a major proportion of the whole order however it is calculated. If ennoblements since 1600 are taken into account, the proportion of ennoblements rises to something like two-thirds. Only a small minority of the nobility, therefore, had noble lineages going back more than a few generations. Far from a caste of feudal remnants, the nobility was an order that was constantly absorbing the richest and most enterprising members of the third estate, renewing in the process its wealth, its energies, and its genetic stock.

A group so well established felt little need to defend itself, for it had no rivals. It is true that the thought of the Enlightenment was full of implications that could scarcely be reconciled with many of the attitudes and assumptions associated with the notion of nobility.[14] It is true, too, that many instances can be found of hostile remarks in the literature of the eighteenth century. No institution so prominent could fail to attract cricitism. But there is no evidence that hostility to the whole principle of the nobility, and its pretensions, was a major preoccupation of eighteenth-century thinkers. Indeed, what is often forgotten is that several of them produced arguments that the nobility was useful, and there could be no higher commendation in 'enlightened' eyes. Noble landowners who tended their estates and resisted the temptation to squander their money in the fleshpots of Paris—argued the elder Mirabeau—were among the most valuable members of society.[15] And Montesquieu suggested that, although the honour so beloved of nobles was really nothing more than a prejudice, nobilities performed an essential public service as the intermediary bodies which prevented monarchs from becoming despots. Nobility, he declared, 'enters in some way into the essence of monarchy, whose fundamental maxim is: *no monarch, no nobility; no nobility, no monarch*'.[16] These arguments still enjoyed wide currency in the 1780s, although by then full-blown attacks on nobility and all that it was taken to stand for were beginning to appear. Figaro's famous soliloquy denouncing Count Almaviva in Act V of Beaumarchais's *Marriage of Figaro* was the talk of Paris when the play secured a performance, after several years of prohibition, in 1784. The younger Mirabeau's *Considérations sur l'ordre de Cincinnatus*, which appeared the same year as part of the debate on the American Revolution, was, as Benjamin Franklin put it, 'a covered satire against noblesse in general',[17] and remained controversial for several years. But the author of *The Marriage of Figaro* was himself a self-styled nobleman, and Mirabeau was a genuine one of distinguished extraction. It was only through influential noble patronage that the play secured a performance at all, and nobles flocked to see it. Provincial parlements, it is true, tried to ban it for its subversive opinions; but most nobles who took an interest in such things did not feel in the least threatened. Why should they? In a country where competition for ennobling offices sent their prices soaring,[18] and where even men who were soon to put themselves at the head of an anti-aristocratic revolution—men like Brissot, Danton, Marat, and Robespierre—were decorating themselves with spurious nobility, there was evidently little serious danger to the position of nobles from the opinion of the educated public.

Nor was there much more from the government. The pre-revolutionary nobility has often been depicted as resolute and united in defence of its privileges, against a government bent on destroying them. But the government never adopted such a programme, lurching as it did from one crisis and makeshift expedient to another. A government staffed by noblemen never dreamed of questioning the social and political pre-eminence of the nobility, or even of making a serious assault on its privileges. If it had mounted such an assault, no power in France could have stopped it, as Louis XIV proved when he destroyed tax-exemption, the most important noble privilege of all. When Turgot proposed to commute the *corvée* into a money tax, and subject the nobility to it, the project became law despite the resistance of the parlement of Paris. It only lapsed, like any other minister's policies, after his fall--which was not the work of the parlement either.[19] When, two years later, Necker introduced his pilot provincial assemblies, with a non-noble majority and voting by head rather than order, nobody was able to prevent him. But all these instances were exceptions to the rule that in general the government did nothing to threaten the nobility or its interests. Turgot, whose long-term dreams might have done so, reserved his most vehement criticisms for ventilation in his own private circle of admirers. Not until Calonne denounced the self-interest of what he labelled the 'privileged classes' in order to break the deadlock in the Assembly of Notables in March 1787, did the ministry publicly offer the nation an anti-noble lead. Even then it found little echo for another year and a half.

Even if it had been attacked by more determined adversaries, the nobility, unlike the clergy with its quinquennial assemblies, had no collective means of defending itself. It was no coincidence that the clergy was consistently more successful in maintaining its privileges and prerogatives. There were indeed the parlements, bodies of nobles by definition, who, it has been argued, had become by the late eighteenth century the acknowledged spokesmen for the nobility as a whole, using their powers to block every measure which threatened noble interests.[20] The power of the parlements, however, has been much exaggerated. So has the degree of unity both between and within them. Nor is it at all obvious that they always had a clear or consistent view of where their own, let alone the whole nobility's, essential interests lay. Above all, it is becoming increasingly clear that large numbers of noblemen did not regard the magistrates of the parlements as in any sense their spokesmen. And in their turn, it seems that the magistrates took a fairly unflattering view of other nobles.[21] Mutual antagonisms within the noble order were far from dead or forgotten, in fact. Arguably, they were the nobility's most significant, and ultimately most fateful, characteristic.

Nobles have traditionally been unsurpassed in their capacity for finding good grounds to despise each other. Perhaps the most popular is that of ancestry. Nobles of old extraction looked down upon those of less distinguished lineage, if indeed they regarded them as nobles at all. To the splenetic duc de Saint-Simon, Louis XIV's ministers remained 'vile bourgeoisie', despite their several generations of office-holding forbears and the high titles that Louis granted them. Yet the duke's own ancestry was undistinguished compared with that of families like the Rohans or the Montmorencys. By the eighteenth century, families with a genealogy going back to the middle ages formed only a small proportion of the nobility as a whole, but they undoubtedly enjoyed greater prestige than the rest. In 1732, they were marked out still further from their fellows by the introduction of the 'Honours of the Court', whereby only families with noble ancestry going back beyond 1400 might hunt with the king or be presented to him. Between then and 1789, only 942 families qualified for these 'honours', and of these only 462 were absolutely beyond query.[22] Unless they got into serious financial difficulties–in which case there were plenty of heiresses of less distinguished ancestry willing to advance their own families–families of ancient nobility tended to intermarry, thus reinforcing their social exclusiveness and contempt for outsiders. And this pattern repeated itself all the way down the noble hierarchy, making what appeared from outside to be a monolithic, united order, in reality a series of largely endogamous groups, suspicious and contemptuous of outsiders. At the bottom of the ladder of prestige came the true *anoblis*–those who had acquired nobility through office in their own lifetime, or the children of such people. Very few offices ennobled their holders immediately and completely. The majority required two or even three generations in the same office to ensure the original holder's descendants full and transmissible nobility. At any given time, therefore, a very important segment of the nobility were not full, recognized members of the order; cut off by their ambition from their bourgeois origins, they were not yet accepted, either, by those whom they aspired to join, a defenceless butt for their sarcasms and their disdain.

Only wealth could transcend the gap between old nobles and new, but in doing so it aroused other antagonisms that were even more virulent. Although nobles owned or controlled most of the key sources of wealth in eighteenth-century France, wealth was very unevenly distributed among them.[23] Perhaps 250 families enjoyed revenues in excess of 50,000 *livres* a year—the richest and most ostentatious group in France. Most of them were nobles of long and distinguished lineage; but at least a fifth were financiers and tax-farmers, many of them with still 'unfinished' nobility, and they mingled with the great courtiers on terms of equality which scandalized old-fashioned provincial gentry whose chief or only asset was a long family tree. For some 60 per cent of the nobility had revenues of less than 4,000 *livres*, which dictated a modest, frugal life-style with little scope for the lavish habits that literary tradition ascribed to nobles. Perhaps 20 per cent enjoyed less than 1,000 *livres*, which put them on a par with modestly off peasants, and far below most of the bourgeoisie whom they disdained. Such noble paupers had nothing but mistrust and envy for their richer fellows, especially those in prominent positions. The latter included the magistrates of the sovereign courts of provincial capitals, not rich by Parisian standards perhaps, but able to afford many thousands for their offices, their urban mansions, and their country seats on the choicest estates around the cities they dominated. In the eyes of the *hobereaux*, the impoverished provincial squirearchy, the robe nobility was a class of plutocratic upstarts, with social pretensions not justified by their recent ancestry. Historians know that this view was a caricature, but it was no less influential for that.

At the same time, however, *hobereaux* were at one with the provincial parlementaires in their hostility to the glittering world of the court and capital, which neither, for all the disparity in their respective fortunes, could afford to enter. For without great wealth no nobleman, however distinguished his lineage, or whatever his right to the Honours of the Court, could exist in the extravagant world of Versailles and Paris. And yet only those close to the centre of power had access to the enormous opportunities for further enrichment which it presented. To those who had, more was given. Familiars of the king and queen enjoyed rich sinecures such as court offices or provincial governorships, or drew lavish pensions for no obvious services rendered. They were also able to secure for their relatives, friends, and dependants the pick of the countless ecclesiastical benefices and well-paid jobs in the king's gift. Many a great lord's solvency, in fact, however great his private revenues, depended on his being able to draw lavish subsidies from public funds.[24] Under Louis XVI, it is true, the government became concerned about the concentration of patronage in the hands of an opulent metropolitan oligarchy. The pension list, when the revolutionaries made it public in 1789-90, showed most money going not to the spectacular parasites of the court but to retired officers in sums of a few hundred *livres*. Ecclesiastical benefices were set aside specifically for the poor nobility. Most striking of all, in the notorious (and much misinterpreted) Ségur ordinance of 1781, an attempt was made to confine the recruitment of army officers to those with at least four generations of nobility behind them. The object was to exclude rich *anoblis* more interested in buying themselves a military background than seriously exercising the profession of arms.[25] But all these gestures served only to highlight the antagonisms between noble and noble rather than to diminish them. The Ségur ordinance, for example, would have excluded most members of robe families from military careers while leaving them open to the dillettante playboys of the court.[26] It only served, therefore, to accentuate the hostility between old and new nobles, rich and poor nobles, robe and sword nobles, and court and country nobles, that were already the bane of aristocratic life. And it also, of course, gave great offence to socially ambitious bourgeois, who were now deprived of the hope of seeing even their great-grandchildren qualify as officers in the most socially prestigious of all professions.

The nobility, finally, was divided culturally.[27] Although there was little explicitly anti-noble thought in the eighteenth century, the majority of nobles probably regarded the world of the intellect with suspicion. Faithful and unquestioning sons of the church, they saw no reason to doubt their confessors' advice that contemporary literature was full of atheism and dangerous opinions, and they certainly saw no compelling reason why they should investigate these matters for themselves. When they had the power, as the magistrates of the parlements did, they used it throughout the century to prohibit plays and condemn books that reflected adversely on the established religion or the social order. But most nobles, wrapped up in their own daily affairs, and unable to afford an education that would sharpen their minds, were probably scarcely aware that they lived in a century of unprecedented literary activity. Their ignorance—or indifference—was more than made up for, however, by an influential minority who were passionately interested in the intellectual life of the day, and were in fact one of its chief driving forces. This minority was concentrated overwhelmingly in the capital, around the great *salons*, the academies, and the major publishers, which together constituted the forcing-house of intellectual life. The richest nobles, such as Lafayette or the duc de Liancourt, were heavily represented, for they could easily afford the education and the leisure to pursue speculative interests, and to dabble in fashionable associations like masonic lodges. But a leaven of philosopher-nobles was also scattered throughout the provinces, where they ran local academies and other intellectual societies, and set up the most prestigious masonic lodges. The parlements may have been, as Voltaire constantly complained, the greatest friends to superstition, intolerance, and repression. But few of them were without their minority of philosophers, liberals, and freethinkers, and in Montesquieu the parlement of Bordeaux produced the most influential political writer of the century.

For a group which constituted such a tiny proportion of the nation, in fact, the nobility played a disproportionately large role in its cultural and intellectual life; and in so doing they helped to elaborate many of the ideas which were to inspire the revolutionaries of 1789 and the ensuing decade. In 1788, this 'liberal' minority, many of whom had had their minds opened by what they had seen while serving against the British in America, viewed the collapse of the old government as an opportunity to introduce reforms and innovations that they had been talking and dreaming about for years. Now was the occasion, they believed, for all men of enlightened outlook to come together in order to regenerate the nation according to sound, modern principles, yet without the evils of despotism. Naturally they saw themselves as the leaders of this movement, as they had led the intellectual world throughout the century. It scarcely occurred to them, in their euphoria at the defeat of despotism during that eventful summer, that the majority of their fellow nobles had no faith either in their leadership or their ideas. Most nobles saw the opportunity of 1788 not as one for laying new foundations, but for settling old scores.

7. The Bourgeoisie

The nobility and the clergy were the only groups in society whose limits were clearly defined in law. Historians have no such easy guidelines to follow when dealing with the bourgeoisie. Paris and other major cities did have their own legally defined bourgeoisie, although the criteria differed from city to city.[1] This status, akin to that of freeman of an English borough or city, was a privileged one and conferred important advantages, such as exemption from the *taille* and various other public burdens; and just to confuse matters, many nobles enjoyed the technical status of bourgeois in the towns where they lived. But nobody, at the time or since, thought that the bourgeoisie was made up of this privileged minority. Who were the bourgeoisie, then? Perhaps the best way to solve this problem is to state what they were not.

Members of the bourgeoisie were not noble. Whatever else they shared with noblemen—and as we shall see it was an enormous amount—they did not enjoy the same social status. There was indeed a twilight zone between the two, inhabited by office-holders whose nobility was not yet complete. But nobody seriously expected that a family with *noblesse commencée* would fail to complete it and fall back into *roture*; and meanwhile such families enjoyed most of the privileges belonging to full nobles. At the other end of the social scale we cannot be so precise, but perhaps the most fundamental criterion was indicated by a sympathetic noble observer who, early in the Revolution, defined the bourgeoisie as 'that entire class of men who live on wealth acquired from the profits of a skill or productive trade which they have accumulated themselves or inherited from their parents; . . . those . . . who have an income which is not dependent upon the work of their own hands'.[2] This meant that the bourgeoisie were not peasants either. Some historians, it is true, have dubbed the richer peasantry a 'rural bourgeoisie' on the grounds that their methods of land management were capitalistic;[3] but the only bourgeoisie whom contemporary peasants would have recognized in the countryside were lawyers, land-agents, or townsmen attempting to buttress their social pretensions by buying rural property. Rich peasants were deeply hostile to all these categories. Bourgeois, of course, originally meant simply 'town dweller', but if bourgeois by definition did not work with their hands, then they were set apart from the bulk of the urban populace, too. The bourgeoisie, therefore, were the non-noble comfortably off, living mostly in towns—though within these wide bounds there was a range and a variety of circumstances which made the subdivisions within the nobility seem minor.

It is not surprising that a group so difficult to define should be equally difficult to enumerate. In the present state of knowledge no estimate of the bourgeoisie's numbers can be much more than an educated guess. Perhaps the most credible recent estimate, commanding confidence through the caution with which it is advanced, is 2,300,000 people, or upwards of 8.4 per cent of the population, in 1789.[4] The same authority suggests that in 1700 the bourgeoisie had numbered no more than 700,000 to 800,000. This means that, over a century in which the number of nobles had remained fairly static, and the population as a whole had only risen by about a quarter the bourgeoisie had almost trebled in size. In this numerical sense, at least, there is little room for doubt that the bourgeoisie was a rising group.

It is as yet impossible to say whether the bourgeoisie's share of national wealth had expanded at the same rate. The only area in which there was clearly a massive growth was overseas trade, an area of activity that was practically a bourgeois monopoly. Wealth deriving from industry, banking, and finance certainly multiplied, too, though hardly at the same rate; but here the bourgeoisie shared the profits with the nobility, in proportions that have never been estimated. In any case, all these sectors of economic life only accounted for about 20 per cent of French private wealth in the 1780s;[5] so that even if they had all been totally in bourgeois hands, alone they could not have made the bourgeoisie the dominant economic group. Of the 'proprietary' wealth that made up the remaining 80 per cent, the bourgeoisie owned somewhat less than 25 per cent of that derived from land, easily the most important element.[6] They owned 47,000 of the 51,000 venal offices, although the capital value of the ennobling offices they did not own probably accounted for well over half the total amount invested in venality. They also probably owned most of the capital invested in government stock (*rentes*), although this is a guess. The most difficult problem in reaching a true picture of bourgeois wealth and its expansion over the century is that it seldom remained bourgeois. Bourgeois riches were constantly being transformed into noble ones through the ennoblement of their owners; so that if, despite this, the bourgeoisie in 1789 was richer than it had been in 1700, the true rate at which it had accumulated wealth must have been phenomenal, and far greater than static figures for any given date might suggest.

Here, then, was an apparently dynamic section of society, with every reason to be brimming with self-confidence. It certainly was by the end of 1789. Most of the evidence from earlier years, however, suggests that this was not the case under the old monarchy. The pre-revolutionary bourgeoisie, in a word, had no class consciousness. They did not see themselves as a distinct social group with its own interests, its own values, and its own way of life which it found superior to those of other groups. They realized, of course, that they were distinct from the mass of those who worked with their hands. They knew that they had risen above the populace. But the value system which told them this was dictated by the group above them, the nobility. The ultimate aspiration of most members of the bourgeoisie was to become noble, and for the most part bourgeois values were more-or-less pale imitations of noble ones. Massive evidence can be accumulated from contemporary literature to document this;[7] but it is also shown by the numbers who entered the nobility over the century. The soaring price of certain ennobling offices in towns where bourgeois capital was abundant suggests that the demand for ennoblement was positively growing,[8] and this is what we would expect if the bourgeoisie was expanding in numbers and wealth. The bourgeois was not content to be what he was; a passenger on the social escalator, he was constantly shedding the traces of his low ancestry, all the time attempting to behave more and more like those he hoped he or his descendants might ultimately join. And of course the constant loss of the richest, the most enterprising, and the most ambitious members of the bourgeoisie into the nobility effectively deprived it, every generation, of the leaders who might help it to formulate a consciousness of its own value.

We should not, however, underestimate the obstacles that stood in the way of any such development. One was the immense parochialism of bourgeois life. Nobles, too, could be deeply parochial, as we have seen, but at least they all shared some sense of belonging to a nation-wide body with certain established privileges and prerogatives, and distinctive values—even if they disagreed about what these values were. For much of the century there is little evidence of such an outlook among the bourgeoisie. And this parochialism was reinforced by the wide variety of circumstances, both geographical and economic, that were to be found among them. How could the 300 or so petty tradesmen and routine office-holders who formed the bourgeoisie of a sleepy little cathedral town like Bayeux feel anything in common with, say, the 2,500-strong bourgeoisie of a major industrial city like Lyons, with its great silk merchants and busy commercial lawyers?[9] To take nearer neighbours, what did the overwhelmingly legal and official bourgeoisie of Toulouse, ten times poorer than the local parlement-dominated nobility, have in common with the millionaire merchant princes of Bordeaux, who could buy and sell many members of their own local parlement and were certainly in a position to marry their daughters to its noble magistrates?[10] But even inside single communities the bourgeoisie seldom showed much unity across its whole range. The differences in wealth even within its sub-groups were often enormous, and in most cases the greatest riches were concentrated in the hands of an exclusive few.[11] In the cities of Languedoc, where Protestantism still survived, religious differences produced deep antagonism among different groups of bourgeois.[12] Most seriously of all, right down the middle of the bourgeoisie ran the second most fundamental of the chasms within pre-revolutionary society. The first was that between manual and non-manual work, the bourgeoisie's lower frontier. The second was that between those who traded for a living and those who did not.[13]

No family, of course, entered the bourgeoisie without passing through trade, because it was the only way to amass enough money to abandon manual work. But equally, nobody stayed in trade longer than he had to, because there was no chance of further social promotion so long as he did. Nobility was the summit of social aspiration, and nobles not only despised trade traditionally, they were also positively forbidden by law to engage in it.[14] Of all bourgeois, therefore, those in trade were the least content to be what they were; and there was never any doubt that sooner or later every bourgeois family would renounce its commercial origins to invest in land, office, or government stock (*rentes*). The only uncertainty was when. Was it better to shake off the stigma at the earliest opportunity, and hope to continue social ascent slowly, through the modest profits of these social investments? Or was it preferable to aim at amassing a really large fortune and buy into a high social level without passing through the intermediary stages? Enough money to buy an ennobling office might accomplish the whole process in one leap, and this explains the high price of the office of *secrétaire du roi* in the affluent western seaports. But the rising price of such offices also suggests that competition for them was increasing; this means that it may have been getting harder for merchants at this high level to acquire such much-coveted emblems of social success. Indeed, since the numbers of the bourgeoisie had expanded over the century, there may have been increasing competition all along the frontier between trade and respectability. After the 1730s, for example, few new venal offices were created at any level;[15] after the partial bankruptcies of 1770 and 1771, *rentes* were perhaps no longer the reliable prospects they had seemed; and the rise in land rents which occurred over the century[16] made it more difficult to lose money as a large landowner and perhaps therefore slowed down the estate market. So social mobility for the commercial bourgeoisie may have been getting more difficult in the 1780s, and if they were aware of this they would have found little reassurance in the moves made by, or on behalf of, various noble-dominated bodies to restrict recruitment to well-established nobles. These measures were not consciously directed against the bourgeoisie. But many bourgeois took them that way, and this was the important thing.[17] Finally, the economic crisis of the late 1780s had inevitable adverse repercussions for everybody involved in trade. They came on top of a series of government measures that had been universally denounced by chambers of commerce up and down the land. In 1784, for instance, the trade of the French West Indies had been opened to all comers, to the fury of the hitherto privileged merchants of Bordeaux and Nantes.[18] Then in 1787, Calonne's commercial treaty with England opened the French market to British hardware and textiles, above all cottons, which were cheaper and of better quality than their French equivalents. The true impact of these measures remains uncertain,[19] but the howls of protest from the merchants and industrialists of Rouen, 'the Manchester of France', are well documented.[20] These were measures for which, unlike harvest failures, the commercial bourgeoisie could blame the established order. But there was little that any of them could do about such grievances, and they knew it. It was only when they were suddenly confronted with the imminent prospect of the Estates-General that they began to think differently.

The non-commercial bourgeoisie can be broadly divided into lawyers, office-holders, and *rentiers*. Other smaller categories, like doctors, could also be included, although a group like teachers belonged mostly to the clergy. Undoubtedly the *rentiers*, 'living nobly' on their revenues and not exercising any profession, enjoyed the most prestige. But all sectors sought to add distinction to their social standing by investing in land, however modestly, whenever they could. In fact, the professional bourgeoisie's pattern of wealth was identical with that of the nobility. Both preferred safe, low-yield, proprietary, non-capitalist forms.[21] And like the nobility, they can be broken down into small, largely endogamous groups in which newcomers were the exception.[22] All recent studies emphasize the provincial bourgeoisie's social, economic, and mental conservatism,[23] and such findings ought not to surprise us. Lawyers, office-holders, magistrates, and buyers of government stock were by definition heavy investors in the status quo. They took their tone and their outlook from the nobility who dominated the existing order of things, and there was no fundamental opposition between the interests of the two. They needed each other. The nobility needed bourgeois agents and lawyers to manage its affairs. It also needed a socially ambitious bourgeoisie from which to recruit new nobles and replenish noble fortunes. The professional bourgeoisie in its turn, cut off from the commerce it had abandoned, needed the nobility to provide it with a set of values and living proof that these values worked.

But this only held true so long as they worked for the bourgeoisie, too, and over the eighteenth century there are signs that they may have been doing so less effectively. *Rentiers*, for instance, were haunted right down to 1789 by the memory of the great crash of 1720, when thousands of them had been ruined. Terray's partial bankruptcies of the early 1770s had demonstrated that it might very well happen again, so that throughout the last two decades of the old order they were jumpy and suspicious, no longer confident in the security of their basic investments. Lawyers, meanwhile, were too numerous for the work available, and while the provincial bar was always dominated by a prosperous few, there were many more who could hardly secure enough briefs to make a decent living.[24] Most, perhaps, accepted this philosophically enough, and in any case many of those nominally qualified as advocates had other sources of income: but increasing competition at the bar[25] must have left its casualties. So must the growing demand for venal offices, reflected in the soaring price for most of them.[26] At the top of the system, ennobling offices were increasingly beyond the reach of most non-commercial bourgeois dreaming of a noble posterity. Only large, fast fortunes made in trade or finance provided the necessary resources. The only people, therefore, who succeeded in ennobling themselves were those who entirely bypassed the conventional hierarchy of respectability, leap-frogging directly from trade into nobility. In so doing, they outraged not only the petty provincial nobility who regarded them as upstarts adulterating the purity of the noble order: they also attracted the envy and hostility of the office-holding bourgeoisie who regarded such rapid social ascent as a mockery of the slow, regular procedures in which they had invested so much capital and expectation. Both they and the petty nobility were disgusted and alarmed by the power of naked wealth to dispense with traditional proprieties and undermine established values.[27] On the eve of the Revolution the bourgeoisie was just as deeply divided between rich and (relatively) poor as was the nobility.

Yet the non-commercial bourgeoisie's attitude to the nobility itself was increasingly ambiguous. On the one hand were all the characteristics of outlook, aspirations, values, and economic interests which bound them together. On the other was a whole range of areas of friction that created resentment and feelings of outraged inferiority. A bourgeois did not need to be very rich to be much better off than many petty noblemen. Such an overlap was obviously anomalous when the price of ennoblement was rising so steeply, and the sense of anomaly was only aggravated by the reaction of poor nobles to the situation. No longer distinguishable from commoners by their wealth, they consoled themselves by laying increasing stress on the one attribute that still did distinguish them, their birth. Distinctions of birth perhaps never mattered more to some than when they were ceasing to matter to most. Meanwhile, the professional bourgeois was no happier than the merchant about the apparent closing of the higher reaches of public office to those without a long noble lineage. Of most concern to the professional bourgeois was the closing up of various parlements, for they were the summit of the judicial structure in whose lower reaches most office-holders were to be found. Never within living memory had most parlements in practice recruited many members from the ranks of lawyers or holders of non-ennobling offices; but a seat on the fleur de lis-covered benches of a sovereign court remained an ideal of which many dreamed, and they did not like to think of their descendants never being able to achieve such a dream.

And these disappointments came on top of constant professional tensions between the parlements and other groups lower down the judicial hierarchy. The *bailliage* and *sénéchaussée* courts, for instance, over which the parlements exercised appellate jurisdiction, were constantly skirmishing with their judicial superiors, but seldom won in the end. Many of their members were very happy to profit from the parlements' discomfiture when they were attacked by Maupeou or Lamoignon, because the reforms of both these ministers offered enhanced status to lower courts.[28] Bourgeois advocates, too, suffered from the haughty attitude of the magistrates before whom they pleaded in the parlements, and bitter clashes occurred.[29] Court magnates or rural squires might cast snobbish aspersions on the nobility of the magistrates of the parlements; but the magistrates themselves had no doubt of it, nor had the bourgeois subordinates who had to bear the brunt of their haughty self-esteem. It was fatally easy in such circumstances to identify these attitudes as typical of all nobles. The fact remains, before 1788 there is little evidence of widespread, conscious, bourgeois hostility to the idea of nobility,[30] or even to the parlements. Indeed, with the exception of a minority of *bailliages* seduced by the prospect of promotion, the whole judicial world united behind the sovereign courts in opposition to Lamoignon's reforms. It was only the prospect of the Estates-General, and the quarrels about its composition, that let the seeds sown over the previous century germinate.

The political crisis provided the sun and rain; but seeds also need fertile soil to grow in. In other words, the minds of the bourgeoisie needed to be prepared for new departures when the crisis broke. After all, they had been the butt of noble disdain for centuries without turning against the nobility. The difference, in 1788/9, was that now the bourgeoisie was well educated. Most of the expanding reading public of the eighteenth century must have been bourgeois; the nobility was not numerous enough to account for the expansion, and the bulk of the populace remained illiterate. This is not to say, as many historians tend to do, that the Enlightenment was a movement for the propagation of essentially bourgeois values. The gospel of rationality, empiricism, and utility could neither have been formulated nor propagated without noble leadership, and its successes were achieved in the face of massive bourgeois indifference. The ideas of the Enlightenment divided the bourgeoisie every bit as much as they divided the nobility, and the conservatism of bourgeois opinion even as late as the spring of 1789 suggests that the partisans of Enlightenment were in a minority outside noble ranks, too.[31] The significance of an unprecedentedly well-educated bourgeoisie did not, therefore, lie in the spread of Enlightenment. It lay rather in an increasing awareness of the importance of national public affairs, an awareness the bourgeoisie shared with the nobility. The public opinion which became such an important political force in the late eighteenth century transcended the traditional social divisions. All leisured groups participated in the growth of an educated, well-informed reading public, used to thinking about and discussing public affairs, and used to being treated as a power worth persuading. The unanimity of this public opinion, and its indifference to traditional social divisions, are shown by the way the bourgeoisie supported the various noble-led movements of opposition to the government between February 1787 and August 1788. They were at one with the nobility in opposing 'despotism', and in the belief that liberal, representative institutions were essential if the men of property who bore the bulk of the direct tax burden were to have any say in the way their money was spent.[32]

What neither noble nor bourgeois members of this well-informed, educated élite gave any thought to, in the heat of the struggle to defeat 'despotism', was the precise forms and procedures that the representative regime to be embodied in the Estates-General was to take. And it was in the struggles over this question, between September 1788 and June 1789, that the bourgeoisie first began to show some consciousness of its own separate character, identity, and value. Better education and wider reading had already done much to break down their notorious parochialism, and give them a sense of belonging to a wider national community. The respect paid by writers, commentators, and the government itself to what the public might think accustomed the bourgeoisie to playing a role in public affairs, even if it was a passive one. Indeed, the role played by the magistrates of the *bailliages*, and the advocates who went on strike against Maupeou's or

Lamoignon's judicial reforms, was more than passive. In all these ways the bourgeois of the 1780s were coming to think of themselves as an integral part of the political nation. It was a notion that would not have occurred to their grandparents, who had merely been silent onlookers of a game reserved strictly for nobles. But the generation of 1789 naturally assumed that, in the new order that the Estates-General was bound to usher in, bourgeois would have an important part to play, and a significant say in what was to be done. Many nobles shared this assumption. But many did not, and it was the challenge offered to bourgeois assumptions by noble conservatives during the preparations for the Estates-General that was to make the French bourgeoisie for the first time realize that it might have separate interests of its own.

8. The Election Campaign, September 1788 to May 1789

At the beginning of September 1788, on the morrow of his triumphant return to office, Necker appeared all powerful. His recall had been celebrated by days of popular demonstrations in Paris; there was an immediate revival of the government's credit which enabled him to cancel Brienne's bankruptcy and start to meet the government's obligations again with freshly borrowed money; he had no credible rivals inside or outside the ministry.[1] But Necker did not regard the unparalleled strength of this position as an opportunity to pursue policy initiatives. He saw himself as a caretaker whose sole task was to maintain stability while France prepared for the meeting of the Estates-General. He seems to have shared the semi-mystical faith so widespread in France during these months that the Estates would solve all problems; and one of his first acts was to hasten that happy day by bringing the date of their meeting forward from 1 May 1789, previously fixed by Brienne, to 1 January. Privately, he felt sympathetic to much of what Brienne had been trying to do; but for the moment he recognized that all the archbishop's policies, good or bad, were irreparably discredited, and that it would be deeply divisive to go ahead with them. Almost all were now abandoned.[2] Above all, the judicial reforms which had inspired such a wave of protest in May and June were revoked, and on 24 September the parlement of Paris returned in triumph to the Palace of Justice. The first governmental act it was called upon to register was that which convoked the Estates-General for 1 January 1789.

What this declaration did not stipulate was how the Estates were to be elected and constituted, and what procedures they were to follow in deliberation and voting. By the *arrêt* of 5 July, Brienne had declared that the government's mind was open on these matters, and he had invited all interested parties to send in their ideas. His main aim appears to have been to gain time, while dividing his opponents by sowing antagonism between the so-called 'privileged orders', on one side, and the mass of commoners, on the other. He was also implying, however, that the government reserved the right to constitute the Estates as it saw fit; and since no precedent existed for a satisfactory constitution of the Estates, the king should simply seize the initiative and dictate one.[3] But this, of course, was exactly what had thrown popular suspicion upon the provincial assemblies that Brienne had tried to establish: bodies chosen according to the government's convenience seemed more likely to reinforce 'despotism' than to check it. Necker's declaration did nothing to dissipate such suspicions, and in many quarters the fear persisted that the government still hoped to avoid the Estates altogether.[4] In these circumstances the battle against despotism seemed not yet to be won, and it was to resolve the uncertainty and deprive the government of a free hand to gerrymander the forthcoming assembly that the parlement attached conditions to its registration of the Estates's convocation on 25 September. In near unanimity it declared that the Estates-General must follow precedent; they must meet in the same way as they had last met, 'according to the forms observed in 1614'.[5]

There is no evidence that the magistrates had thought deeply about the implications of this, or even that all of them knew for certain what the forms of 1614 were. If they had, they would surely have bridled at an organization in which members of the parlements had sat with the third estate rather than the nobility. And there is certainly no reason to conclude that their objective was to ensure that the Estates would be dominated by the clergy and nobility. The parlement had always been the fiercest enemy of clerical involvement in politics, and had spent much of the preceding decade attempting to curb the political ambitions of non-robe nobles.[6] Legal guarantees were what the magistrates sought, against 'ministerial despotism' on the one hand, and against mob-rule 'democracy', on the other. Only the day before, in its very first act upon reassembling, the parlement had forbidden the tumults and demonstrations that had occurred almost daily in Paris since Brienne's downfall a month earlier. It wanted orderly, peaceful elections according to clearly understood and legally unchallengeable rules, ensuring the return of deputies who were serious, responsible, and independent.[7]

Unfortunately, the parlement's own tactics since the end of the Notables had helped to ensure that the time for cool and orderly deliberation was past. An atmosphere of public excitement had been whipped up that was not to be easily calmed. Lamoignon's attempt to muzzle the sovereign courts had provoked a nation-wide furore that dwarfed the earlier reaction against Maupeou, and drew into political activity groups that had never played any significant public role before. The sudden collapse of the government had seemed to vindicate all this activity. Understandably, those who had involved themselves in the protest movement did not now feel disposed to become mere onlookers again. Besides, in many provinces the opposition had not confined itself to negative opposition. It had produced positive reform proposals of its own. The main one, of course, was invariably the convocation of the Estates-General; but close behind it in many provinces came a call for the revival or creation of provincial estates to replace the 'despotism' of the intendants and provide a more representative form of provincial government than Brienne's suspect assemblies. Estates, as various parlements had been arguing for decades, would be independent and have extensive powers over taxation. Only in areas which already had estates but where, as in Artois or Provence, they were little more than ciphers, did Brienne's assemblies arouse enthusiasm, for only there were they obviously more representative in character than the previous order.[8] But in the aftermath of Lamoignon's attack on the parlements these peculiarities attracted no attention. Petitions poured in from all over the country demanding provincial estates. Mostly they were organized and drafted by noblemen meeting together in unauthorized assemblies—a 'noble revolt' which scandalized Louis XVI and left him bitter and resentful towards noble political pretensions throughout the subsequent winter. But the first success was achieved by the movement in Dauphiné, which drew its support and much of its inspiration, too, from circles far wider than the nobility.[9]

The Dauphiné nobility was indeed first in the field with a petition for the revival of the province's estates, defunct since early in the previous century. But on 14 June the three orders of Grenoble, under the bourgeois leadership of the judge, Mounier, and the advocate, Barnave, repeated this call and began moves to convene a meeting representing the whole province to press the demand. Early in July, Brienne made vague promises that the estates would be revived, but this did not prevent deputies from all over the province from meeting at Vizille on the 21st. There they not only reiterated the call for estates; unlike campaigners in other provinces, they made clear demands about how the estates should be constituted. The deputies were all to be elected, rather than sit as of right; they were to vote by head, rather than in their separate orders of clergy, nobility, and third estate; and the third estate was to have as many deputies as the clergy and nobility put together. Whatever taxes the estates might vote, they added, should fall with equal weight on all members of society.

The Vizille assembly was a turning point in two ways. In the first place it proved an inspiration and example to all those who were campaigning for estates in other provinces. On 2 August, the government formally decreed an assembly to deliberate on the revival of the Dauphiné estates, thus vindicating the movement in its defiance of earlier ministerial disapproval. This inspired other provinces to clamour for the same treatment; and the provincial estates movement became a major preoccupation for the nobles of several provinces throughout the autumn. Even more important, however, were the Vizille assembly's demands concerning the form and procedure of the estates. There was nothing new about double representation for the third estate, or about vote by head. Several provincial estates had always had the former, and so had the provincial assemblies introduced between 1778 and 1788; and whereas no previously existing estates voted by head, all the recently instituted assemblies had done so. Nor were totally elective assemblies or equal taxation new ideas. Most existing estates were largely made up of unelected members who sat by right, but none of the assemblies had included such members; and equality of taxation had been unambiguously conceded by the Notables. The new feature was the combination of all these points in a single programme, enunciated by a meeting in which all three orders freely participated. It was, moreover, applicable not merely to the special circumstances of Dauphiné. As discussion began, in September 1788, about the forms, procedures, and concerns of the forthcoming Estates-General, the Vizille programme offered an obvious model. But it was a model that bore no similarity to the 'forms of 1614' demanded by the parlement of Paris. On that occasion the three orders had been equal in size, and had voted as orders.

It was a few days before the point sank in, for even fewer people outside the parlement than inside were aware of what the forms of 1614 actually were. But once they became widely known there was an explosion of hostile publicity in Paris which took the parlement, the government, and a fair proportion of the general public, too, completely by surprise. Generations of historians have viewed this as the moment when the bourgeoisie, seeing the magistrates at last in their true selfish and aristocratic colours, abandoned its former heroes overnight and began to campaign exclusively for its own interests, embodied in a doubled third estate and vote by head.[10] That there was a marked change in the political atmosphere and the preoccupations of pamphleteers is certain; but how far this was the bourgeoisie's work is far less clear. Far from a spontaneous outburst of bourgeois anger against the parlement and the 'privileged orders', what now occurred seems to have been a carefully orchestrated attempt to whip up such anger in a bourgeoisie which as yet had given little thought to electoral questions. And this was largely the work of a political club comprising a majority of nobles which met at the house of Duport, a dissident magistrate of the parlement of Paris itself.[11]

Political clubs had blossomed in Paris during the spring and summer of 1787, and so suspicious was Brienne of their role in fomenting opposition that late in August of that year they were forbidden to meet. But after his fall this order lapsed, clubs proliferated once more, and early in November they were formally authorized. Among the most important was that meeting at Duport's, later called the *Société des Trente*.

In its ranks were magistrates, priests, courtiers, bankers, academicians, lawyers, and journalists—in fact, a complete cross-section of the metropolitan social and intellectual élite. Its history remains shadowy, but its aims were clear enough. Its members believed that the nation was in need of radical regeneration, that the Estates-General would provide the occasion for bringing this about, and that therefore the form and membership of that body were of crucial importance. No progress would be made, they thought, if the Estates were organized on outmoded principles which relegated the bourgeoisie, with all its wealth, talent, and experience, to a subordinate position which it could only resent when recalling how much more Necker or Calonne or even Brienne had been prepared to grant it in the days of despotism. The regenerated nation must eschew invidious distinctions between fellow citizens, and a truly national assembly must fully represent the best elements in the nation. The 'forms of 1614' must therefore be opposed with all available means, while as a first step the number of third estate deputies should be doubled.

And so this 'conspiracy of gentlemen', as Mirabeau called it,[12] launched a campaign of pamphleteering and agitation directed at the bourgeoisie. No doubt Necker's release, early in September, of everybody arrested by the previous ministry for writing or distributing anti-ministerial tracts, placed a number of ready pens at their disposal. The aim was to awaken bourgeois political consciousness, to promote denunciation of the parlement's ruling, and to fan agitation for the doubling of the third. Paris was flooded with broadsides, emissaries were sent to major provincial centres, and model petitions were circulated for local authorities to send in to the government. It was a calculated assault on national public opinion, and once launched it snowballed. By December it had run far beyond the control of the *Trente*, as everybody with a view to express rushed into print. Ultimately it succeeded. But it only did so by setting out to arouse bourgeois resentment against the 'privileged orders'. An inevitable result was to alarm the more timid members of these orders themselves and push them into intransigent defence of their separateness and their privileges. Thus a strategy motivated by a desire to banish social divisions from politics had the opposite effect of making all sides more conscious of them; and the social fusion which had characterized the Vizille assembly and indeed the *Société des Trente* itself became the exception rather than the rule.

This process began almost immediately. Necker, like everybody else, was taken by surprise by the parlement's insistence on the 'forms of 1614'. He did not wish to have his hands tied so early in deciding on the form to be taken by the Estates. In order to gain time, and in order, like Calonne before him, to outweigh the parlement's authority with something even more imposing, he decided to reconvene the Assembly of Notables to discuss the composition of the Estates. Convening on 6 November and dispersing on 12 December, the Notables considered fifty-four questions formulated by Necker.[13] Nobody doubted, however, that the crucial one was the number of third estate deputies. On this the Notables were not unanimous, but only 33 out of 147 voted for doubling their numbers. With that clear, they felt free to be more adventurous on the question of voting. Only 50 rejected voting by head outright—but of course, with the third estate outnumbered, voting by head lost much of its significance. Nor could the reaffirmation of the Notables' earlier commitment to equality of taxation soften the blow. Necker may have hoped that the Notables would bow to and join in the swelling public demand for the doubling of the third, and would thus lend him the authority of their opinion in decreeing it. Instead, the strength of the clamour scared them, they became intransigent, and the onus of decision was thrown back on the government. And recourse to the Notables had also made it impossible to hope to convene the Estates on 1 January. They were now put off once more until the end of April.

The nominally secret proceedings of the second Assembly of Notables were widely leaked and discussed throughout their proceedings, much to the fury of some of them. On their dispersal, five out of the seven princes of the blood published a declaration, previously presented to the king, denouncing the rage of publicity and arguing for the forms of 1614. Such defiance only increased the clamour from what was now calling itself 'national' or 'patriotic' opinion. Pamphlets began to appear by figures who were soon to achieve national importance—men like Sieyès, Volney, Servan, Roederer, or Rabaut de Saint-Étienne—all denouncing the selfishness of the 'privileged', and providing a whole new range of arguments for doubling the third based largely upon the wealth, the talents, the experience, and the tax burden of the bourgeoisie. The leader of the patriotic party in Dauphiné, Mounier, reminded the public that the assembly convoked in his province to discuss the form of provincial estates had recommended doubling the third and vote by head, and that the government had accepted these principles.[14] Petitions began to come in from provinces, cities, and corporations throughout the country calling for double representation. By the end of December they numbered 800 and were still arriving. Often, it is true, they were forced upon reluctant local authorities by well-organized local pressure groups taking their lead from Paris; but this was certainly not the case everywhere, and the importance of the metropolitan lead diminished, while that of local initiative grew, as the movement gained momentum.[15] On 5 December, even the parlement of Paris, whose announcements had touched off the clamour, trimmed to the wind. After a careful campaign by Duport and a number of liberal magistrates, it declared by a narrow majority that it had never intended by calling for the 'forms of 1614' to prescribe the numbers or procedures of the Estates-General; it had merely meant that the electoral units should be *bailliages* and *sénéchaussées*.[16] There were no binding precedents to settle the questions of the number of deputies and of voting. These were matters for the king and his ministers to decide. Although 'patriots' were quick to note that the parlement had still not declared itself favourable to the doubling of the third, it had at least dissociated itself from the inflexibility of the Notables.[17] Its defection left them totally isolated. The provincial nobles and clergy who might have given them a solid basis of nation-wide support had remained silent throughout the autumn, or had thrown themselves into local controversies over provincial estates rather than national questions. All this made it easy for Necker to ignore the

Notables' opinions, and to follow the popular course to which his instincts always led him. By the *Result of the King's Council of State of 27 December* it was formally decreed that the number of third estate deputies in the Estates-General should be doubled so as to equal that of the clergy and nobility combined.

It was an enormous triumph for the campaign launched three months earlier by the *Trente*, but this did not mean that 'patriotic' opinion was now satisfied. The *Result of the Council* also specified that third estate electors were free to choose deputies belonging to the other two orders, which aroused fears that deferential bourgeois electors would not choose candidates of their own kind. Above all, vote by head was not decreed, but merely permitted if the orders in the Estates-General, once convened, agreed to it. Nobody now expected that the nobility and clergy would agree to any such thing; and so throughout January 1789, denunciation of the 'privileged orders', their separation, and their political ambitions continued. It was in this month that Sieyès published perhaps the most famous pamphlet of the Revolution, *What Is the Third Estate?*, in which he argued that so long as the nobility and clergy refused to share common rights and burdens with their fellow citizens they were not part of the nation and so should enjoy no rights of any sort. It was in this month, too, that the Swiss journalist Mallet du Pan made his much-quoted observation summarizing what had happened since the previous September: 'The public debate has changed. Now the King, despotism, the constitution are merely secondary: it is a war between the Third Estate and the two other orders.'[18]

These antagonisms had been much exacerbated in the provinces by controversies over the revival of provincial estates. The harmony achieved on this question in Dauphiné proved quite exceptional. In Guienne, the noble sponsors of a scheme for estates proposed to double the third but maintain vote by order, thereby losing it the support of 'patriotic' leaders; nor could the scheme's sponsors agree among themselves about the area the estates were to represent.[19] Controversy about the revival of the estates of Provence went back to the spring of 1787, and when Brienne convoked them instead of a provincial assembly early in 1788, they were denounced as unrepresentative. The example of Dauphiné later in the year inspired renewed calls for total reorganization from a coalition of unrepresented bourgeois and nobles excluded from sitting because they owned no fiefs.[20] Similar resentments arose in Franche Comté, where estates defunct since 1666 finally met in November 1788 with most of the province's nobility excluded by the shortness of their ancestry or lack of qualifying fiefs. This both tore the nobility apart and provoked strident denunciation by third estate pamphleteers.[21] In Artois the estates had never lapsed, but here, too, three-quarters of the nobility were excluded by regulations demanding several generations of noble blood and ownership of certain lordships. Their meeting in January 1789 was almost entirely taken up with quarrels over this issue.[22] The nobility of turbulent Brittany, by contrast, remained fairly united, for every full Breton noble could sit by right in the estates, which had remained a power in the province throughout the century. But this domination by the 'iron swords' was challenged throughout the autumn of 1788 by the 'patriots' of Rennes and the rich commercial centre of Nantes, who called for limits on the number of nobles, free election of third estate deputies, and an increase in their number to equal that of nobility and clergy combined. In the last days of 1788, the Breton estates convened under the traditional rules, but clashes between the orders on the issue of reform led to their suspension on 3 January. Later in that month there was street fighting in Rennes between law students calling for reform of the estates and servants of the conservative nobility.[23]

All these clashes, and others like them elsewhere, could only further polarize opinion throughout the provinces on the issues of the composition of the Estates-General and the rights and privileges of the nobility. They also provoked second thoughts about the whole idea of provincial estates, which had seemed to be the pattern of the future only a few months before. The Vizille assembly of July 1788 had expected that provincial estates would choose the deputies to the Estates-General from each province, and this assumption had lent importance to all discussion about estates. Whoever controlled particular estates would ultimately dictate the character of the national ones. But on 24 January 1789, the government decreed that deputies should be elected, in four-fifths of France, by assemblies of each order in the *bailliages* and *sénéchaussées*. Only in Dauphiné itself were the estates to have any role at all. The movement for estates, already flagging as the public became disillusioned by the difficulties it raised, was completely destroyed by this regulation. Those who continued to espouse it were brushed aside with the assurance that the Estates-General, in its wisdom, would settle such matters. The national elections now became everybody's primary concern.

The electoral arrangements, a mixture of precedent and *ad hoc* provisions elaborated between January and April 1789, were extremely complex, and exceptions to the general rules were numerous.[24] In most of the country, however, the procedures followed certain broad principles. Every electoral district was to choose deputies for all three orders, but each order was to elect its own deputies and deliberate and vote apart. All bishops and parish clergy in each *bailliage* could attend the assemblies of the first estate in person, but chapters and monasteries could only send representatives. All full nobles were also allowed to take part in the second order's deliberations in person. This was, of course, impossible in the case of the third estate. Here, every male taxpayer over twenty-five was allowed to participate in a primary assembly, which chose two delegates for every hundred households in its area to form part of the third estate electoral assembly in each *bailliage*. Thus clergy and nobility elected their deputies directly, the third estate indirectly. Nor were the duties of the assemblies confined to electing. They were also authorized to draw up *cahiers*, lists of grievances and reform proposals, for the consideration of the Estates-General and the guidance of their deputies; and eventually each *bailliage* would produce three composite general *cahiers*, one for each order.

No time scale was prescribed for the elections, other than that they should be complete before the date on which the Estates were to convene at Versailles. In the event, the last deputies were elected almost three months after that date, so the elections ran from February through to July. Most, however, took place in the course of March and April, against a background of economic crisis, popular unrest, and continuing public effervescence and pamphleteering. Yet apart from promulgating the regulations themselves, the government made hardly any attempt to influence the outcome. There were no government candidates, official or unofficial, and no serious efforts to prevent the election of undesirables. Necker, still paralysed by that quasi-religious faith in the potential of the Estates which served as a substitute for planning and thought, maintained the neutral stance he had adopted on his return to office; and there could be no government candidates when the government itself stood for nothing. This absence of a guiding hand, which no other power was certainly in any position to provide, left the outcome of the elections to be determined very largely by local circumstances. But one feature applied everywhere. Everybody expected the Estates-General to bring change. Nobody expected it merely to reinforce an old order which had already proved unviable. The very institution of the *cahiers* encouraged such expectations, for why draw up lists of grievances if there was no prospect of redressing them? The peremptory, mandatory tone of many of the *cahiers* suggests that their drafters did not seriously question whether the Estates would do what they required. All this meant that those with little desire for change played no great part in the elections. The lead was taken, in all three orders, by people who wished to alter the way things had been done before 1787. The Revolution was to be, as one contemporary later described it, everybody's revenge.[25] The elections of 1789 demonstrated this for the first time.

The very composition of the electoral assemblies of the first estate, the clergy, was revolutionary. Deliberative bodies were nothing new to a Gallican church which nominally governed itself through quinquennial assemblies, and, at diocesan level, by elective *bureaux*. But the parish clergy had always been unrepresented on these bodies. The last thirty years of the old order had witnessed widespread signs of unrest among them over the injustices of their subordinate position, but all attempts to concert their action in order to achieve reforms had been thwarted by the authorities. The electoral regulations of 1789 offered them the upper hand for the first time. In the electoral assemblies they completely outnumbered their bishops and the representatives of chapters and regular orders combined. Throughout the spring, too, there were plenty of pamphleteers to remind them that their interests were the same as those of the third estate from which most of them were recruited. The bishops, canons, and monks who had always excluded them from church government and impropriated the tithes that were rightly theirs, were in contrast either nobles or bound to the nobility in their interests. 'Why talk of three orders of citizens?' wrote one. 'Two suffice; two alone are justified by experience; everyone is enlisted under one of two banners—nobility and commons. . . . Like the country itself, the clergy is divided. . . . The parson is a man of the people.'[26] That hostility to the bishops and non-pastoral clergy in general was a guiding principle among clerical electors is evident from the results of the elections. Eventually 303 clerical deputies arrived at Versailles in May 1789, and of these 192 were parish priests and only 51 were bishops.[27] Many of the electoral assemblies witnessed bitter clashes between leaders of the parish clergy and representatives of canons and regulars resentful at the way these former clerical underlings had outmanoeuvred and outvoted them.[28] Eventually, in June, the very deputies elected to the Estates were to split among themselves, an event which heralded the final merger of the three orders.

But, despite the antagonisms so evident from the beginning of the electoral process, this was no foregone conclusion. There were plenty of *bailliages* where no clashes occurred between rival clerical groups, and the *cahiers* of the clergy were muted on the grievances of parish priests, and emphatic on a number of issues that united all ecclesiastics. They were unanimous, for example, that Catholicism should remain the established religion and retain control of the educational system. They were suspicious and uncertain about tolerating Protestants, and reluctant to surrender the power to censor the dangerous and godless publications of the 'impious and audacious sect which adorns its false wisdom with the name of philosophy and which works to overthrow the altar'.[29]

They were prepared to give up their fiscal privileges. These points of agreement were just as significant as calls, deriving largely from the parish clergy, for open access to bishoprics and canonries, an end to pluralism and non-residence, diocesan government by elected synods, an increase in clerical stipends, and an end to tithe impropriation. The parish clergy expected the Estates-General to redress a whole range of their grievances against the great corporation of which they were the humblest but arguably the most useful members. They did not expect it to challenge the central role which the church and its ministers played in national life; indeed, they expected it positively to reinforce that role. Much of the subsequent tragic history of Catholicism during the Revolution derived from the failure of laymen to take full account of these ambiguities.

The electoral assemblies of the clergy seem positively harmonious in comparison with those of the nobility. Even before they convened there was the vexed question of who qualified to sit in them. In provinces like Artois, Franche Comté, and Provence, great bitterness had already been aroused by the attempts of old-established families to monopolize the provincial estates and so the choice of deputies to the Estates-General. These exclusivists in their turn were incensed by Necker's decision that the elections should be by *bailliage* and that all nobles, irrespective of the length of their ancestry, should have a vote. The nobles of Brittany, spurned in their demand to have their deputies elected by the provincial estates, voted to boycott the elections altogether and sent no deputies to the Estates-General. Several *bailliages* were so seriously divided that rival factions elected their own deputies and drew up their own *cahiers*, and the Estates-General had to choose between them. And everywhere, only nobles with full and transmissible nobility received a summons to the electoral assemblies. This excluded several thousand *anoblis*, members of rich and ambitious families who had paid good money to escape from the third estate and now found themselves brusquely pushed back into it. And once in session, the noble assemblies rapidly fell to quarrelling among themselves, venting long-harboured resentments against fellow-nobles as often as they addressed themselves to the momentous questions now confronting all Frenchmen. Much grumbling was directed at the courtiers and Parisian socialites who suddenly appeared in distant provinces where they owned lands expecting to be elected automatically by awe-struck *hobereaux*. Many of them, like Lafayette, had an unexpectedly difficult struggle—although in the event the metropolitan aristocracy still enjoyed enough prestige to secure far more seats than their numbers alone seemed to warrant.[30] The nobility of the robe was not so lucky. Whereas in some districts, like Bordeaux, it was able to outmanoeuvre a group bent on excluding magistrates from the Estates, elsewhere it mustered little support. In Franche Comté it was the magistrates who were outmanoeuvred.[31] In the end only 22 members of the parlements were sent by noble electors to Versailles, outnumbered by 'sword' nobles by 8 to 1.[32] The noble elections were a triumph for the natural but hitherto silent majority among the nobility—provincial, poor, relatively inarticulate, politically inexperienced, but determined to use this unlooked-for opportunity to disavow those who had formerly usurped the role of their spokesmen. Not surprisingly, it was a conservative majority, and the anti-noble agitation of the preceding six months had only stiffened its conservatism. Yet the political and social intransigence of the nobility in 1789 has been much misinterpreted and exaggerated. Although there were at most 90 liberals, more or less favourable to the claims of the third estate, out of 282 nobles sitting at Versailles in May 1789, their only clear common features were that they tended to be younger than their conservative fellows, and to have had urban rather than rural upbringings. Surprisingly, there were plenty of metropolitan conservatives and provincial liberals.[33] Most surprising of all is the degree to which the nobility as a whole was prepared to consider the claims of the third estate sympathetically. This emerges clearly from the noble *cahiers*.[34] On the key question of whether voting in the Estates-General should be by order or by head, only 41 per cent insisted on vote by order; and whereas only 8 per cent called for vote by head pure and simple, the rest were prepared to envisage it in certain circumstances. No less than 89 per cent of noble *cahiers* favoured the abandonment of the nobility's fiscal privileges, privileges which indeed few nobles had sought to defend ever since the Assembly of Notables had agreed to give them up in 1787. Some 40 per cent condemned privileged access to the courts of law (*committimus*), 24 per cent believed that ennoblement should result from merit and talent, and less than 10 per cent of all demands were for the preservation or reinforcement of noble distinctions. Almost a quarter of the *cahiers* actually called for nobles to be allowed to trade or pursue other supposedly degrading professions. While none of this suggests a positive enthusiasm to accede to all that the third estate might demand, it does indicate that the mood of the assembled nobility in 1789 had not yet solidified into total hostility, and that nobles were still open to persuasion of the need for further concessions. Besides, on constitutional matters there were vast areas of common ground between nobles and commoners. If, despite all this, deputies of the second and third estates were slow to come together in May 1789, it was owing to procedural difficulties and political circumstances within the Estates-General once they met, at least as much as to genuine disagreements about fundamental principles.

The third estate elections did not always proceed uneventfully,[35] but on the whole they were far more harmonious than those of the other two. Despite the opportunities offered by the widest franchise Europe had yet seen for influencing the choice of deputies and voicing hitherto smothered grievances in *cahiers*, many voters abstained from the primary assemblies, notably in the south. Even when they did not, they tended to send to the electoral assemblies delegates who were literate, educated, and had little in common with those who chose them. The majority of third estate electors were peasants, and, in the towns, artisans; but the assemblies which actually elected the deputies who went to Versailles were overwhelmingly bourgeois in character. So then, inevitably, were the deputies themselves.[36] There were no peasants and no artisans. Considering their wealth and importance within the pre-revolutionary bourgeoisie, there were relatively few men of business: 76 merchants, 8 manufacturers, and 1 banker only made up 13 per cent of the third estate deputies in all. These elections were a landslide victory for the non-commercial, professional, and proprietary bourgeoisie, and if they are to be assigned any social significance at all, these results look like a vote against the power and the possessors of naked wealth. In contrast, between 10 and 11 per cent of deputies described themselves as some sort of landowner, and the true proportion of landowners must have been far higher. About 25 per cent were lawyers—advocates and notaries. And no less than 43 per cent, 278 deputies, were office-holders. No doubt these proportions reflect to some extent the balance of discontent within the bourgeoisie. But on the whole, we do not need to elaborate sweeping social or economic reasons for the emergence of these men as the third estate's deputies.[37] Lawyers and office-holders had always been the natural leaders among non-noble Frenchmen. They were the best educated, most used to public speaking, to drafting public documents like *cahiers*, and to handling the legal complexities of public business in the courts. Naturally, then, they dominated the electoral assemblies, and naturally they got elected. The arrangements for the third estate elections, for example, were in the hands of the magistrates of the *bailliage* and *sénéchaussée* courts, and no less than one-fifth of the third estate deputies came from these courts, making one-twentieth of their entire personnel in France.[38]

The impression that the third estate deputies were not sent to Versailles by a bourgeoisie burning with resentment at the injustices of its position is borne out by an analysis of the third estate *cahiers*. These were more abundant than those of the other two estates, since most primary assemblies produced *cahiers* as well as the secondary ones which actually elected deputies. But the grievances of peasants and artisans were largely strained out of the general *cahiers* which became the final official statement of third estate opinion. This did not necessarily mean that the general *cahiers* were less radical. The electoral process seems in fact to have clarified the political consciousness of those taking part, and the longer they were involved, the clearer and more radical their ideas became.[39] The widespread circulation of model *cahiers* by the patriotic clubs of Paris also played an important part in this process.[40] Even so, the radical tone of the general *cahiers* of the third estate, the first collective expression of the political opinions of the eighteenth-century bourgeoisie, has been much exaggerated. It is true that they called overwhelmingly for vote by head in the Estates-General, for equality of taxation and careers open to the talents, but it would have been astonishing if they had not after the barrage of propaganda to which the public had been subjected on these issues since the previous September. They also called, in common with the other two orders, for regular meetings of the Estates-General, no taxation without their consent, guarantees of civil liberty, and the establishment of reformed provincial estates. But the mandate for abolishing seignorial rights, venal offices, or trade guilds was far from overwhelming, while that for the confiscation of church lands and the abolition of tithes, urban and provincial privileges, monasticism, and nobility itself was very weak indeed.[41] The desire seems to have been to reform and improve existing institutions rather than to destroy them root and branch. Yet within a year most of these things *were* destroyed, and by an assembly composed of the men elected in the spring of 1789. What turned them from reformers into radicals was the same thing that split the clergy and soured the nobility: not the experience of the old regime, but that of the months of May, June, and July 1789.

A reading of the *cahiers* of all three orders together suggests that the comments of contemporary observers about a war between the privileged orders and the rest of the nation ought to be treated with some caution. In the last analysis, only one major issue divided the first two estates from the third—that of vote by head. Even on this the desire of the clergy and nobility to preserve vote by order was not as firm and absolute as propagandists for the third, then and since, have assumed. On fiscal equality the third estate was pushing against an open door, as it largely was, too, on careers open to the talents. Aside from the question of voting, the most striking thing about the *cahiers* is the wide measure of agreement they display about how France should develop in the immediate future. Everybody agreed that the age of despotism was over. Everybody agreed that the king should rule through a constitution, in which laws would be made and taxes voted only by national consent expressed through regular meetings of the Estates-General. There was wide agreement that government should be representative at all levels, and that the provinces should each be run by estates. Nobody disputed that individual rights must in future be guaranteed by law, and only some members of the clergy were opposed to full freedom of faith, thought, and expression. Everybody agreed that the church needed extensive reform and that the status of the pastoral clergy ought to be raised. Everybody agreed, too, that the economy should be stimulated by reforms such as the abolition of internal customs barriers and the standardization of weights and measures.

Behind the arguments over voting in the Estates, in fact, there lay a broad reformist, liberal consensus which transcended all three orders. Once they had finally united, in July 1789, the way was clear for this consensus to become the basis for the National Assembly's reforming activity until 1791. But before that could happen the great outstanding obstacle had to be cleared away, and this proved more difficult than anybody had imagined. In the process, suspicions, resentments, and antagonisms were aroused that the idealistic *cahier* drafters of the spring could not have dreamed of. The uniting of the orders, indeed, was only achieved by the intervention of an element which had no place in the liberal consensus—the common people. Nor, having accepted popular assistance in order to save the National Assembly at the very moment of its creation, did the revolutionaries of 1789 have any intention of allowing the common people any permanent place in the new order, for which only men of property would qualify. But in order to calm the populace they had to make major concessions, and willy-nilly these carried what was to become the ideology of the French Revolution far beyond the demands and expectations of the electors of 1789.

9. The Economic Crisis

Down to the spring of 1789, the forces pushing France towards revolution were almost entirely political. There was no underlying social crisis; it seems unlikely that such social discontents as surfaced in the *cahiers* would have done so without the stimulus provided by the constitutional wranglings that followed the collapse of the government. Nor did economic factors play an important part. Senior administrators had foreseen budgetary difficulties in 1786 and 1787 as long before as 1783, and although a number of the government's bankers were in difficulties in 1787, difficulties caused partly at least by economic uncertainties, these problems seem to have subsided by the time of the governmental collapse in August 1788.[1] By then, however, it was already possible to foresee major economic troubles in the following spring. By mid-August it was clear to peasants all over France, as they surveyed cornfields ravaged by persistent drought followed by massive storms, that the harvest of 1788 was going to be disastrous. This accident of nature was to have incalculable consequences for the history of France, and the world. It did not cause the French Revolution, but it did dictate the sort of revolution it turned out to be.

Every economy in eighteenth-century Europe was dominated by the agricultural sector, and France's was no exception. Agriculture, in its turn, was dominated by the need to produce cereals for the population's staple diet. In normal times French agriculture had little problem in coping with the demand. The very fact that the population had increased from 20 millions at the beginning of the century to over 28 millions in 1789, shows that there was spare capacity to meet growing needs.[2] Nor did the eighteenth century witness any of the general harvest failures that had led to catastrophic falls in population in the seventeenth century. The last of these occurred in 1709. After that, although serious regional and inter-regional shortages continued to occur, there were no great mortalities as a result of them. Maritime provinces were able to import supplementary supplies from abroad, and a first-rate system of roads enabled afflicted inland provinces to be helped from those where there was abundance. The government regarded the assurance of adequate grain supplies as one of its prime duties; and when it abandoned controls on the grain trade it did so in the genuine belief that a free market would supply needs better than the traditional system of close supervision. Few consumers, however, appreciated the sophisticated economic arguments behind these *laissez-faire* policies, and it was to prove unfortunate for their ministerial proponents that they were first attempted at a time when harvests were becoming less predictable.

The years from the late 1730s until the late 1760s were bumper ones for French agriculture. Production, prices, and rents rose steadily, and substantial regional and year-to-year fluctuations did little to moderate the general trend. But around 1770, the fluctuations became more violent, and they remained that way for the next two decades.[3] These years were not without good harvests, for instance, in 1779, 1783, and 1787; but in 1770, 1774, 1778, 1782, 1784, and 1786 there were serious deficiencies affecting several provinces at once, and in 1788, for the first time within living memory, the harvest was poor to catastrophic almost everywhere. A major natural disaster, therefore, came as the culmination of two decades of difficulties and uncertainties.

Poor harvests do not always spell disaster for growers. Scarcity drives up prices and producers can often make better profits than in years of abundance. But the effect of the uncertain harvests of the 1770s and 1780s, averaged out, was to depress agricultural revenues. Nor were cereals the only crops to suffer. In 1778, the vintage failed completely, depriving many small cultivators of an important supplement to their revenues. Subsequent years were abundant, and by the mid-1780s there was a wine glut. These wild fluctuations deprived the market of all stability. In 1785 and 1786, there was also a general shortage of hay, making it impossible to obtain adequate fodder for livestock at reasonable prices, and thereby forcing sales of cattle and sheep in an oversupplied market.

Only one thing remained steady and inexorable throughout these see-saw years, and that was the rise in rents. A steadily increasing population meant that land was always in short supply. In addition, there was a particularly marked rise in rents in the 1770s, when landlords who had granted leases in the booming 1760s, not foreseeing the profits their tenants would make, sought to compensate themselves with exceptionally hefty increases. Yet this was just when the boom conditions which might have allowed tenants to bear them easily were ending.

Landlords, then, were cushioned from the agricultural depression, and so were large-scale producers who could stock up in abundant years and hope to profit from these stocks in lean ones. But large-scale producers were the exception. France was a land of small peasant cultivators who had no safety margins. Those with a surplus needed to sell it to survive, and those without seldom grew enough to guarantee even self-sufficiency. This was why the wine crop was so important. Grapes have a high yield per cultivated area, and all over France peasants supplemented their slender livelihood with vines grown on odd patches of marginal land. The chaos of the 1780s in the wine market devastated this resource. Peasants also earned extra revenue by cottage industry; indeed, in a period when factories were unknown and urban workshops small and largely devoted to luxury products needing a high level of skill and training, peasants constituted the vast majority of the ordinary industrial labour force. Unfortunately, industry, too, was seriously affected by the economic crisis.

The most important industry in France, accounting for half of all industrial production, was textiles. And after food, clothing was the largest item in most people's expenditure. Accordingly, when food prices rose or merely became more erratic, less was spent on clothing and demand for textiles fell. 'Bread has become so dear', wrote the intendant of Rouen, France's cotton capital, in 1768, 'that the labourers, artisans and workers . . . have much trouble in earning enough to buy it in order to live and feed their families. The price of bread seesaws with the workshops, they fall into torpor when bread is too dear . . . the makers and manufacturers are loaded with unsold goods.'[4] And so, the wild ups and downs of the harvests in the 1770s and 1780s were reflected in equally spectacular variations in the textile trades born of unstable demand. The effect on weaving towns like Rouen, Amiens, Nîmes, or Lyons was devastating. Weavers tended to be laid off at exactly the times when the price of bread was already straining their limited wages. And in the countryside peasants impoverished by the agricultural crisis found that they could not rely on weaving in order to make up for their losses. The conditions which brought the failure of fodder crops in the mid-1780s also affected the supply of flax and hemp upon which low-grade textile production relied; in 1787 the silk harvest failed; and finally the commercial treaty of 1786, which came into operation in May 1787, opened the French market to competition from the cheap, high-quality products of the industrial north of England. So the textile industries were already reeling when the 1788 harvest disaster brought a drop in demand sharper than anybody could remember. Production may have fallen by as much as 50 per cent in a few months, leading to massive unemployment or at best, for those lucky enough to stay employed, substantial wage reductions.

Wages had not kept pace with inflation over the eighteenth century. To a generation that has lived through the 1970s, the 62 per cent by which the cost of living rose between the 1730s and 1789, or even the 45 per cent by which it had risen since 1770, do not seem as alarming as they once did. Yet the fact that wages had only risen by 22 per cent represented a serious long-term decline in living standards for all wage earners—which included large segments of the peasantry. And behind these bland figures lie concealed massive year-to-year fluctuations in the cost of living which wage rates took no account of. Indeed, as we have seen, the only way wages moved in times of difficulty, if they moved at all, was down. In normal times, it has been estimated,[5] a wage earner spent 50 per cent of his income on bread, and a further 16 per cent on other food and drink. When harvests were bad, and the price of bread went up, other items had to be sacrificed. In Paris the price of bread began to rise on 17 August 1788, the day after the government went bankrupt.

A 4-pound loaf costing 9 *sous* went up to 9½ *sous*, amid some popular grumbling. By 7 September it had reached 11 *sous*, two months later 12 *sous*, 14 *sous* by Christmas, and 14½ *sous* by February.[6] Meanwhile it was a lucky workman, assuming he still had a job at all, who earned more than 20 *sous* a day. During the worst days of the spring of 1789, in fact, bread alone was absorbing up to 88 per cent of an average Parisian worker's wage. And then on top of everything else came a prolonged and bitterly cold winter. Roads were blocked with snow and rivers froze, immobilizing mills, paralysing the transportation of fuel, throwing even more workers out of employment. Large stretches of the Rhône and Loire valleys were then flooded when at last a thaw set in. 'The wretchedness of the poor people during this inclement season', reported the British ambassador, 'surpasses all description.'[7]

The only sector of the economy not in crisis at the end of the 1780s was overseas trade. It had expanded fivefold over the century and was still expanding. By the 1780s, its volume equalled that of England, and although one of its foundations was the export of textiles and manufactures to southern Europe and the Spanish colonial empire, the most important and fastest-growing element was largely independent of the fluctuations in the domestic French economy.[8] This was the re-export of colonial products to northern Europe, which formed the basis of the spectacular prosperity enjoyed by ports like Bordeaux and Nantes.[9] The growth in overseas trade was not without its vicissitudes, especially during wartime when the British fleet cut off the sources of colonial wealth in the French West Indies. The abrogation in 1784 of most of the regulations confining the trade of these islands to French shipping created much consternation in the western ports, exposing them to stiff competition from North America. The glut in wine supplies in the 1780s was a matter of particular worry to the merchants of Bordeaux, who on this account were one of the few commercial groups to see advantage in the Anglo-French treaty of 1786. But none of these problems appeared fundamental. If the economic crisis of 1788/9 affected overseas trade at all, it was only to produce a deceleration in the growth rate, not a downturn. In 1789, the trade of Bordeaux, which alone accounted for 40 per cent of French overseas commerce, reached its greatest volume of the whole century. It was only external factors—the Saint-Domingue slave revolt of 1791 and its aftermath, followed two years later by renewed war against Great Britain—that was to bring this unique trading boom to an end.

In any case, spectacular though this boom was, it was peripheral. Only the inhabitants of a few ports and their rural hinterlands derived much benefit, and the greatest profiteers were as often as not foreign immigrants. The vast bulk of Frenchmen were untouched by the prosperity of overseas commerce, and it was no solace to them in 1789 that one sector at least of the economy was relatively immune to the economic crisis. None of the others were. And this meant that very few of Louis XVI's subjects did not have their discontents sharpened by economic difficulties in the spring of 1789, at the very moment when they were being invited to list their grievances in the *cahiers*. The economic crisis compounded the excitement whipped up by the political one, and the atmosphere of these months, full of suspicion towards everything emanating from the government, made it even more tempting than usual to blame economic misfortunes on the policies of the king's wicked advisers. Nor was it as if the economic ideas which guided these advisers, at least down to August 1788, had not got the government into trouble before. Calonne's chief economic counsellor was Dupont de Nemours, the last notable survivor of the physiocratic clique who had been so influential in the 1760s and mid-1770s. Physiocrats believed that the chief object of economic policy was to boost agricultural profits so as to promote productivity, and this was what Dupont hoped the Anglo-French treaty of 1786 would do. In exchange for admitting English industrial goods, France hoped to break into a rich new market for her agricultural surpluses. What in fact happened was that England took very little from France, her advanced agriculture being easily able to supply all her needs except wines. Even here the English continued to prefer port. France, on the other hand, was flooded with English hardware, woollens, and cottons; and the outclassed, outpriced spinners and weavers of Rouen, Lyons, Amiens, and other textile centres blamed all their troubles on the treaty, as their *cahiers* make clear.[10]

Even less opportune was the decision of Calonne, maintained by Brienne, to lift all restrictions on the grain trade, a physiocratic experiment that had first been tried in the 1760s. Then, in combination with a series of mediocre harvests, free market conditions had produced uncertain supplies and these in turn bred popular disturbances.[11] In 1769, Terray reimposed controls. In 1774, Turgot removed them once more, but a poor harvest pushed Paris bread prices up to 14 *sous* a loaf by the following May. By then riots were sweeping the market towns of northern France, the populace were arbitrarily fixing a 'just' price for grain, and this 'flour war' reached the capital itself as bread prices peaked.[12] The riots, the last substantial ones in Paris before 1787, were put down by armed force and Turgot continued with his policy. He even extended it to the wine trade. But when he fell he was succeeded by Necker, already a known advocate of controls, and they were periodically resurrected until 1787. Their abandonment in that year coincided with an abundant harvest, and this period of a free market contributed nothing to the dearth of 1788/9, apart, perhaps, from depleting reserves through exports. But it certainly aroused memories of hardship on previous occasions, when controls had been abandoned; and nothing did more to win Necker popular support, when he returned to office in August 1788, than his announcement that grain supplies were once more to be controlled. Such a gesture could have little practical effect in the face of the immensity of the harvest failure, but it did reassure the populace that the minister was determined not to abandon them. They were to show their gratitude in July 1789 by not abandoning him.

Meanwhile, as their purchasing power collapsed, the populace looked desperately for ways out. They ate less bread. They economized on clothing, heating, and lighting, particularly agonizing steps during that terrible winter. They also began to talk, ominously, of not paying rents, dues, and taxes, and here and there during the spring these ideas were carried into practice. But the first instinct of a hungry populace was, as always, to take direct action to control grain supplies and fix the price of bread. Throughout the spring a series of grain riots swept the country.[13] The first serious looting of bakers' shops occurred in Brittany in January. Here the rioters who made the conflicts over the Breton estates so turbulent were partially motivated by apprehensions about the bread supply. The worst disturbances, however, took place in March and April. During March, there were attacks on bakeries and granaries in Flanders, while in Provence mobs forced local authorities to fix low prices for bread. Early in April, the women of Besançon led a bout of popular price-fixing, and later in the month there were disturbances over grain in Dauphiné, Languedoc, and Guienne. In several of these areas unrest continued into May, and by then it had also made its appearance in Paris itself.

The capital had been jumpy ever since bread prices had begun to rise at the end of the summer, and worries over this question no doubt played their part in the demonstrations which greeted the fall of Brienne and the recall of Necker. Tension certainly rose with the price of bread and the approach of the dreaded *soudure*—the gap between the exhaustion of the previous year's grain stocks around the end of May and the next harvest in August. Yet despite the frenetic publicity and political agitation which stirred the capital far more than anywhere else throughout the winter, calm was maintained, and when rural grain riots reached the surrounding countryside, they did not spread into the city as they had in 1775. When an explosion came, it happened quite unexpectedly, and indeed in circumstances that could hardly have occurred before 1789. The Paris elections to the Estates-General took place later than in the rest of the country, and under different rules. The electoral assemblies did not begin to meet until 20 April, and on the 23rd, as that for the third estate of the Sainte Marguerite district was deliberating upon its *cahier*, the prosperous wallpaper manufacturer Reveillon was heard to remark that steps should be taken by the Estates to bring down the cost of bread so that wages could be reduced to 15 *sous* a day.[14] Reveillon, whose factory, with its 350-strong work-force, was one of the biggest in Paris, was a generous and enlightened employer. During the winter's depression he had even paid his unemployed operatives 15 *sous* a day relief. But even to mention wage reductions, however guardedly, when bread was so expensive and likely to become more so, was imprudent. Over the next few days angry crowds assembled all over eastern Paris as rumours about wage reductions spread, and on the 27th a mob sacked the house of Henriot, a saltpetre manufacturer who had made similar remarks to Reveillon at another electoral assembly. By now troops were on the streets in order to keep order, but on the 28th they were overwhelmed by a mob several thousand strong which sacked Reveillon's house and factory shouting *Vive le Roi! Vive Necker! Vive le tiers état!* They were only dispersed by the arrival of fresh troops who killed or wounded at least fifty of them and perhaps many more.

The Reveillon riots brought Paris face to face with the problem of public order, which had been haunting the provinces for months. And they demonstrated that the government was uncertain of how to deal with it. It is true that the troubles were eventually quelled, as they had been in 1775, by the ultimate guarantee of law and order, the army.[15] But Necker was not sure whether the troops would remain reliable, and this may help to explain why more were not sent in to prevent further trouble at the outset. The army was certainly stretched that spring. Most units were being used for internal police work—guarding grain convoys, supervising markets, quelling disturbances. They were tired by the constant activity, and their morale suffered from being deployed in small, isolated detachments. But once deployed, they dealt with rioters briskly enough, and everything suggests that they would have obeyed even harsher orders just as promptly. What they craved was leadership, and firm support from their superiors when they faced hungry mobs of desperate rioters. Instead, they served a minister who did not trust them, in containing a populace whose adulation he craved. The French Guards, who had done most to put down the Reveillon riot, were blamed by the authorities for allowing it to get out of control in the first place. It was not a response calculated to strengthen the soldier's resolve as he faced crowds appealing to him not to fire on his hungry fellow citizens. Officers knew that their troops could not withstand such pressures indefinitely, and they, too, began to doubt in the course of June whether they could rely on their men's obedience. These doubts only reinforced the government's hesitations about the use of military force.

The Estates-General, therefore, were elected, and met, against a background of riot, disorder, and popular anxiety which left few parts of the country untouched. It was a situation which imposed enormous pressures on the deputies arriving at Versailles during the last week in April and the first in May. They could not be unaware of the boundless popular hopes and expectations invested in them. Two years of agitation, and then the invitation to draw up *cahiers*, had led the ordinary people of France to expect prompt, comprehensive action to solve their problems. One of the anxieties expressed by the Reveillon rioters was that the meeting of the Estates, originally scheduled for 27 April, had been postponed for a week. But any detached observer could have seen that popular worries and grievances were far too numerous, diverse, and conflicting to admit of easy solutions. Besides, the deputies had their own priorities which were often completely at variance with those of the populace. What worried them most, men of wealth and property that they were, was chaos. All around them law and order seemed to be collapsing, and the forces of government seemed unable, or unwilling, to do much about it. There was no knowing where popular violence would lead unless it was contained; the Reveillon rioters may have cheered for the third estate, but some of them also shouted 'Down with the rich!' It began to seem that the only way to prevent the people from taking the law into their own hands was for others to do so first. Already in March the electors of Marseilles had supplanted the official city authorities and set up a 'citizens' militia', which they hoped would be more effective in maintaining public order. Several other towns followed the example in the course of April and May. The deputies arrived at Versailles with no such radical plans in mind, but they were conscious that firm and vigorous action was urgently needed if they were ever to get down to the task of national regeneration which all agreed to be the Estates-General's true purpose.

10. The Estates-General, May and June 1789

The Estates-General finally convened at Versailles on 4 May, and almost at once euphoria gave way to frustration and disappointment. The government still had no lead to offer. Neither the king, nor the keeper of the seals, nor Necker, who all made speeches at the opening session, proposed any programme. They all dwelt on the problem of the finances, and expressed pious hopes that the Estates would help to elaborate a new system. They conceded that the moment might be opportune for discussing reforms in education, the freedom of the press, and the law. But they all denounced the spirit of hasty innovation, and they showed no inclination to satisfy the clamour for deliberating and voting by head which had become the keystone of the 'patriotic' programme during the spring. Nobody condemned the idea, but Necker made it clear that he was not prepared to impose it as he had the doubling of the third the previous December. Sometimes deliberation in common would be appropriate, he declared, sometimes not, and it certainly would be inappropriate at least until after the nobility and clergy had freely renounced their fiscal privileges. Accordingly, the orders were instructed to separate and verify their credentials.

Everybody seems to have been disappointed by the timidity of the government's approach, and its reticence on the question of voting placed the deputies of the third estate in a particular dilemma. If they were to verify their credentials separately, they would be implicitly accepting the legitimacy of separate deliberation and voting. If they refused, it would be an act of open revolt. Deputies from Dauphiné and Brittany, with a record of twelve months' more-or-less successful defiance of authority behind them, were undeterred by such fears, and called for the three orders to verify their credentials in common. Most other deputies appear to have been unsure what to do, and in these circumstances, although discussions were protracted, the views of those who were used to working together, and who knew what they wanted, prevailed. The third estate refused to constitute itself into a formal body, or to transact any official business pending the reply of the other two orders to an invitation to verify credentials in common.

The reply of the nobility was prompt. On 7 May, the question was put to the vote, and by 188 votes to 46 it was rejected.[1] On 11 May, the nobility declared itself 'constituted' and ready for business. In taking this line the nobility was not, as many historians have perhaps over-hastily concluded, declaring itself against all further discussion of common deliberation. The subsequent participation of noble commissioners in conciliatory talks shows that their attitude was not irrevocably settled. They merely meant that they accepted Necker's analysis of the proper course, and that if privileges were to be lost they should be freely renounced rather than taken away. Many also felt in honour bound to observe the mandates in their *cahiers* to reject, or at least to resist, vote by head, which seemed the obvious corollary of common verification. No doubt this is why the 'liberal' minority did not dissociate themselves from their order's decision, however much they regretted it. They knew they had lost a battle, but not the war. Among those who voted with the majority a good 40 were not entirely inflexible,[2] and with time they and perhaps others might be prepared to change their attitude. Unfortunately, the longer the liberals remained with their colleagues and refused to break ranks, even though their objective was to persuade the majority to be more accommodating, the more they became identified in the minds of onlookers with that majority, and the easier it became for commoners to believe that all nobles in their hearts were against real reform, and prepared to obstruct the work of national regeneration for their own selfish motives.[3]

The attitude of the clergy was less clear-cut. They, too, voted to reject common verification, but only by 133 to 114. They certainly did not go on to constitute themselves formally, as the nobility had; and they proposed that each order should nominate commissioners to discuss the matter further. This proposal proved acceptable to the other two orders, and between 23 and 27 May there was a series of conciliatory talks. The auspices for agreement were not improved, however, by the rejection on 14 May by the third (or 'commons' as they had now begun to call themselves) of a motion committing them to 'respect the property, rights, and prerogatives of the two upper orders'; and there is ample evidence that throughout the period of the conference their suspicions about the nobility's true motives were growing rapidly.[4] None of this reassured the nobility, nor did the inflammatory reports which certain third estate deputies were beginning to send home to their constituents with talk of stalemate, inaction, and obstruction. The talks proved fruitless, and on the 27th the nobility passed a resolution declaring that the 'constitution', meaning the separation of the orders, was the only safeguard of liberty. In the light of this it is all the more surprising that all three parties agreed on the 28th to a royal proposal for further talks in the presence of ministers. Clearly there was a genuine desire still for some way ahead among deputies throughout the orders. But the proposal that eventually emerged from the renewed talks—that each order should verify its own credentials, but that contested ones should be settled in a plenary session of all three—was rejected by the nobility. This saved the third from doing the same. By 9 June, the conciliatory conferences had petered out. After a month of activity the Estates-General, in which so much hope and expectation had been placed, had achieved nothing.

This does not mean that nothing had happened. The third estate had made clear that it was only prepared to transact business in a general assembly voting by head. Leaders had emerged from among a body of deputies who on 4 May had been largely unknown; and the hesitant mass had been led into a policy by the Dauphinois, whose commitment to deliberating and voting in common went back to the Vizille Assembly of a year earlier, and the Bretons, whose hostility to noble pretensions had been forged against the uncompromising egotism of their own provincial aristocracy. In their turn, the ferocious noises emanating from some third estate deputies had scared the nobility, most of whom were probably more timid than reactionary. Afraid to support 'the 46', as the liberals were known, and suspicious of the sort of people they were, politically inexperienced country squires were easy prey for a handful of determined and eloquent conservatives. And the clergy, meanwhile, were torn between the rapidly polarizing extremes of nobles and 'commons'. They were obviously split far more seriously and far more evenly than the nobility. Third estate leaders were aware that many parish priests, as well as a few leading bishops, were tempted by their repeated (and carefully calculated) appeals to join in common verification of credentials for the sake of peace, harmony, and progress. This month of deadlock and inaction also produced changes at court, where Necker no longer reigned supreme. His apparent inability to control the disorders of the spring, followed by his seeming tacit complicity in what was seen as the obstructive and insubordinate attitude of the third estate, had produced a whispering campaign among alarmed courtiers. Its burden was that Necker was allowing the kingdom to drift into chaos. The queen, the king's brothers, and certain of Necker's ministerial colleagues began to talk of emergency steps to stop the rot; and without the knowledge of the hitherto all-powerful minister, the number of troops in the Versailles region was quietly increased.[5] But the public did not fail to notice this buildup, or the foreign regiments who accounted for the greater part of it, and rumours began to circulate that there was an aristocratic plot to obstruct the work of the Estates so as to give the king an excuse for dissolving them. Finally, throughout this time the price of bread had continued to rise, and with it, popular panic. Riots and disorders had abated somewhat in the expectant atmosphere of the first few weeks of the Estates, but when nothing immediate was done they resumed. Now the rioters not only blamed the government for their troubles, as they always had. The causes of their misfortunes were now also held to be the 'privileged orders', who seemed to be preventing the arrival of those better times that the meeting of the Estates had led everybody to expect.

This was the situation when the third estate, as a national body, took its first overtly revolutionary step. On 10 June, Sieyès, author of *What Is the Third Estate?* and the leading ideological mentor of the 'commons', moved that a final invitation should be sent to the nobility and clergy to verify credentials in common, and that the process of verification should begin at once and proceed whether members of the other two orders accepted the invitation or not. It was a motion to ignore their separate existence, and it was passed by the overwhelming majority of 493 to 41.

Up until this moment, throughout two years of struggle and debate, opponents of the government had never questioned its legitimacy, but only the legitimacy of some of the things it tried to do. They had sought to influence or restrict the way it used its power, but not to usurp that power for themselves. Even the deadlock of the first month of the Estates had been caused by a prolonged attempt to secure action by agreement under existing conventions rather than by unilateral action. But the motion of 10 June 'cut the cable', as Sieyès put it. By it, the deputies who made up the third estate notified their intention to dictate terms. They were not the first to lay claim to power in 1789. The electors of Marseilles, who in April had swept aside the established authorities in a bid to control popular disorder, had that distinction. But those who passed the motion of 10 June did so as the elected representatives of 95 per cent of the French nation, and they believed that the nation had given them both the right and the duty to take control of its affairs. On this date, then, the bourgeoisie became revolutionary; and the transfer of power which lay at the heart of the French Revolution began.

Strictly speaking, the study of the Revolution's origins should end here, when the event itself begins and the revolutionaries seize the initiative. But if this initiative had failed there would have been no French Revolution, or at least not the one that actually occurred; and over the next few weeks there were serious attempts to make it fail. Not until these were defeated was the Revolution guaranteed a future. Nor was it clear by 10 June 1789 what the revolutionaries' programme was to be once they had taken power. Almost all their deliberations up to that time had been about the question of voting or the mounting problem of public order. The revolutionaries' programme only emerged finally in August, and its form could not have been entirely predicted in June, since much of it was deeply influenced by the events of the intervening weeks. Not until those events have been analysed, therefore, can the origins of the Revolution be fully understood.

Between 12 and 17 June, in accordance with the resolution of the 10th there was a roll-call of all deputies—clerical, noble, and third—elected by each *bailliage*. The government ignored this development, and the nobility conducted a leisurely discussion of the third's invitation. But the clergy was thrown into turmoil by it, and there was a week of impassioned debate culminating in a vote, of disputed validity, to join what was called the 'general assembly' to check election returns. This motion was carried by 149 deputies, not all of whom were parish clergy feeling at one with the third estate from which they had sprung. Bishops and clergy alike were genuinely divided over where they thought the church's best interests lay, and it is clear that those who voted for common verification did not share the third estate's view that this was tantamount to supporting vote by head on every issue.[6] Yet by this time the intentions of the third were perfectly obvious, and a handful of clergy had committed themselves to them. Not waiting for their order's vote, on 13 June three priests broke ranks and presented themselves at the roll-call. Over the next few days another sixteen, not all parish priests, joined them; and they were present and did not dissent when, on 17 June, the assembly adopted a title. The one it chose, after rejecting various proposals designed to leave doors open to the other two orders, was National Assembly. Only eighty-nine deputies voted against this bold solution, and none of them refused to accept the result of the vote. To emphasize that it was a sovereign body, the Assembly went on to pass a resolution provisionally authorizing all existing taxes, and guaranteeing the national debt. There could have been no clearer indication that the deputies were claiming supreme power in the state; and although these motions were carried to shouts of *Vive le Roi!* their plain implication was that the king was no longer sovereign in France.

In the face of such a direct challenge to his authority even Louis XVI was at last spurred into action. Over the preceding fortnight he had been more than usually inert owing to the death of his eldest son on 4 June, which had prostrated the whole royal family with grief. But on 18 June, it was obvious that he must act urgently before matters got completely beyond control. It was decided to hold a 'Royal Session' of the Estates-General on 23 June,[7] at which the king would propose a programme. Until that time, the hall where the third estate had been meeting was closed. But no official notice of this closure was given. The first the deputies knew about it was when they arrived as usual on the morning of 20 June. Outraged and suspicious, they reassembled in a nearby indoor tennis court and there took an oath never to disperse until they had given France a constitution. It was one more assertion that they were subject to no other power in France; and unlike the motion of 17 June transforming them into a National Assembly, it was practically unanimous. It was in fact framed by the Dauphinois leader, Mounier, who had led the resistance to unilateral action on 17 June. When the Assembly next met, on 22 June, it was joined by the 149 clerical deputies who had voted for common verification three days earlier and, even more significant, three noble deputies from Dauphiné. The orders, therefore, had already begun to merge by the time of the royal session on the 23rd, and the revolutionary deputies were more united than they had ever been when they appeared at it.

The idea of this session appears to have been Necker's.[8] His aim was to have the king promulgate a reform programme so extensive that the radicals would be overwhelmed and forget their wranglings about voting. He planned to have the king announce that the orders would deliberate and vote in common on all important issues, though retaining their separate identities; and, to salve the tender consciences of nobles and clergy who felt bound by their *cahiers* to keep to the separation of the orders, binding mandates were to be cancelled. The events of 10 to 17 June would be ignored, but the regular future meeting of the Estates was to be guaranteed; they were to consent to all new taxes, and they were to co-operate in the elaboration of a more equitable tax system, a new structure of local government through provincial estates, guarantees for individual liberty and press freedom, careers open to the talents, and a number of other specific reforms. But between 19 and 23 June, the king came under heavy pressure from the queen and his brothers to modify these proposals, and they were so changed by the 23rd that Necker ostentatiously stayed away from the royal session and sent a letter of resignation.

Almost all the specific reforms envisaged by Necker remained in the proposals announced at the royal session. But the king began by quashing the resolutions of 10 and 17 June and reaffirming that the separation of the orders was sacrosanct. They were authorized to deliberate in common, but the nobility and clergy were accorded an effective veto on all proposals affecting themselves. The king ended with an implied threat that if he did not receive co-operation he would press on with the reforms alone. As these provisions were unveiled, the deputies of the nobility and some of the clergy did not disguise their exaltation. The third received them in sullen silence. Proposals whose generosity might have overwhelmed them at the beginning of May, many of which, indeed, were largely to be adopted by the National Assembly in subsequent months, left them indifferent when included in a package which destroyed all they had done during the preceding fortnight. The crown's attempt to give a lead came too late, and all too obviously under pressure of circumstances. So when the king left, with an order to the deputies to disperse and reconvene the next day in their respective chambers, the third estate refused to move. The king, informed of this, took no steps to force them, absorbed as he now was in a desperate attempt to persuade Necker to withdraw his resignation. Everybody had noticed the minister's absence, rumours flew about that he had been dismissed, and Versailles was flooded with ugly crowds from Paris. After an elaborate show of authority, therefore, the king caved in at the first sign of resistance; and as Necker yielded to royal entreaties to stay on in the interests of public tranquility, the National Assembly voted to reaffirm the tennis court oath and the provisional authorization of taxes. Aware now, too, of a growing threat of force, it voted that the person of deputies should be inviolable. After that it proceeded with its deliberations as if nothing had happened.

The royal session, then, had failed. The king had thrown away his authority almost as soon as he had tried to reassert it. The deputies of the clergy and the nobility were quick to see that the royal support promised on 23 June would not materialize. Accordingly, on the 24th, most of the clergy who had not joined the Assembly on the 22nd now did so. Others followed on the 25th, and on this day, too, forty-seven liberal nobles finally abandoned the conservative majority of their order to join the third estate in sympathy with whom they had consistently voted. Defiance of the royal orders was becoming general, and every day now Versailles was swamped with crowds cheering Necker and threatening those who persisted in not joining the National Assembly. On the 24th, the archbishop of Paris, one of the few remaining clerical separatists, was almost lynched, and the doors of the clerical and noble meeting places were picketed by jeering spectators. The only hope the king now had of retrieving the situation lay in the army. There were already 4,000 troops stationed around Paris, and on 26 June orders were issued to bring up 4,800 more. But it would take time for them to concentrate, as well as to work out plans for their use, and popular disturbances in both Versailles and Paris were growing daily more violent. The French Guards, hub of the metropolitan garrison, whose morale had steadily deteriorated since the Reveillon riots, were no longer reliable. Whole units were passing resolutions not to obey their officers, and they took pride in having rescued the archbishop without firing a single shot at their fellow citizens. There was a rumour, which seems to have been believed at the palace, that 40,000 Parisians were ready to march on Versailles if the orders did not unite. All this meant that the king had to temporize until his forces were stronger. On 27 June, therefore, he sent orders to the rump of the clergy and the separatist noble majority to join the National Assembly. The clerics obeyed instantly, the nobles after a brief discussion in which it rapidly became clear that they had no alternative. In this way seven weeks of stalemate were at last brought to an end, and the Estates-General finally became a

National Assembly recognizing no separate orders in the state, and voting by head. That night there were wild scenes of jubilation in Paris, while at Versailles the king was cheered, there were fireworks, and the whole town lit its windows. 'The whole business now seems over', wrote Arthur Young, that ubiquitous English observer, 'and the revolution complete.'[9] The revolutionary action inaugurated on 10 June certainly seemed to have been brought to a successful conclusion, and the National Assembly appeared to have seized sovereign power. But Young was not prepared to believe that the court would 'sit to have their hands tied behind them', and of course he was right. The troops ordered up on 26 June were on the move, and on 1 July a further 11,500 received marching orders. By the end of that week nobody doubted that the king was still prepared to use force to bring the Revolution to an end. The only thing that could prevent him was counter-force, and as yet the Assembly had none at its disposal. It was saved only by the people of Paris.

11. The People of Paris

There were popular disturbances all over France during the spring and early summer of 1789. The deputies to the Estates-General were all well aware that the country appeared to be dissolving into anarchy, and everybody agreed that once the question of voting was settled, the problem of public order ought to be the first priority. The anguished feeling that they were fiddling while Rome burned did much to embitter the deputies of the third estate, and many of those of the clergy, during the deadlock over verification of credentials. Yet it is unlikely that, even with the whole country disturbed, resolute deployment of troops could not have maintained order until the harvest was in. It is equally unlikely that the disturbances would have decisively influenced the course of political events if they had not spread to the capital. Arthur Young, arriving in Nancy on 15 July, before the news of the fall of the Bastille, was told, 'We are a provincial town, we must wait to see what is done at Paris; but everything is to be feared from the people, because bread is so dear, they are half-starved, and are consequently ready for commotion.' And yet, commented Young, 'they dare not stir; they dare not even have an opinion of their own till they know what Paris thinks . . . . Without Paris I question whether the present revolution, which is fast working in France, could possibly have an origin.'[1]

Paris in 1789 was a city of perhaps 600,000 or 650,000 inhabitants—six times larger than the largest provincial towns. Within this population, side by side and often quite literally one on top of another in the tall apartment blocks which provided the bulk of the city's accommodation, were the most enormous contrasts of wealth and power to be found in France. With the exception of a few shippers and the occasional provincial financier, all the richest people in France lived in the capital. So did most of those involved in political power. It is true that the seat of government was at Versailles, twelve miles to the west, but the courtiers who occupied the cramped and foetid apartments of the royal palace did not regard it as their home. They spent most of their time, and took most of their pleasures, in spacious and elegant mansions in the capital city. Paris remained the centre of social life, and about 20 per cent of the population were rich enough to sample at least some of its pleasures. They can be broken down into perhaps 5,000 nobles, 10,000 clergy and 105,000 bourgeois, who among them employed at least 16 per cent of the rest of the population to wait on them as domestic servants. The really rich, of course, were far less numerous. Only 2 to 3 per cent of marriage contracts concluded in 1749, for instance, were worth over 100,000 *livres.*[2] But fortunes of this size were not necessary to live well in Paris; besides, over the century most of the rich grew richer thanks to the rise in rents from which they derived most of their income. The wages of the poor, in contrast, only rose at one-third of the rate of inflation; nor was this the only way in which rich and poor were growing further apart. Although many of the comfortably off still occupied the traditional first floors of apartment blocks whose tenants got poorer the higher they lived, whole new smart quarters were being built from the 1760s onwards, where the wealthy and fashionable tended to congregate. Central districts like the Marais, where the great had lived earlier in the century and under Louis XIV, were forsaken in favour of spacious new suburbs out to the west, or redeveloped districts to the south of Saint-Germain des Prés. So the poor were increasingly segregated in the eastern suburbs both north and south of the river, on the two islands at its very centre, and in new districts of cheap lodging houses to the north. They saw the rich only when they came clattering through in their carriages, often at breakneck speed despite the narrowness of the streets. Accidents were frequent, and they caused enormous popular resentment since the victims were invariably those who were too poor to buy or hire carriages themselves.[3] It is not surprising, then, that 'the rich', however vaguely defined, should attract the sort of hostility that burst out during the Reveillon riots, and was to keep resurfacing throughout the Revolution.

For the vast majority of the capital's population was poor. In 1749, 39 per cent of those who married made no contract, and so can be presumed to have had no property worth recording. Indeed, possibly 100,000 of the metropolitan population at any given time are practically unrecorded in any of the sources used by historians, for they constituted a 'floating' population of migrant workers who drifted in and out of the city in search of work when there was none in the rural areas they came from. Sometimes they settled permanently in Paris if they found regular work. Perhaps two-thirds of the city's population, in fact, were not born there, for deaths constantly exceeded births in the crowded and insanitary conditions under which most of the working population lived. Only by constant and massive immigration did Paris maintain its population at all. Yet large numbers of immigrants came and went with the seasons, like the Limousin masons and construction workers who returned home every winter to plough the fields and live off the harvest. Only in times of rural distress was the rhythmic quality of immigration disrupted. Then, thousands of extra peasants flooded in in search of work which as often as not was hard to find in the city, too. Thus in 1789 as many as 30,000 more immigrants than usual may have been present in Paris as a result of the economic crisis, and few of them, thanks to that same crisis, had much hope of finding work. There were 22,000 people alone who were employed in government work-creation projects (*ateliers de charité*), and 14,000 more filled the capital's hospitals and poor houses. Most of these penurious immigrants lived in conditions of great squalor, huddled together many to a room in garrets or cellars, fraternizing only with fellow immigrants from their own regions, unknown to the distributors of local charity and therefore unlikely to benefit from it.[4] When employment failed they turned to begging, to crime, and if they were women, to prostitution. And the authorities feared them. They, and men of property generally, believed that the rootless, drifting, anonymous migrant was the cause of most public disorders in the capital. After all, he had nothing to lose. He was also presumably open to bribery from ambitious politicians wishing to foment trouble for their own purposes. There was, for instance, a very widespread belief that the duc d'Orléans had paid such people to foment the Reveillon riots.

In fact, there was little to fear from the 'floating' population. They were not prominent in the Reveillon riots or any subsequent disturbances. Because they had nothing to lose, they had nothing to defend, and therefore lacked strong motivation to take to the streets. Dangerous individually, as potential criminals, they lacked organization, the habit of acting together, and strong common interests to push them into purposeful collective action.[5] These qualities were much more likely to be found among the settled Parisian population (even if of provincial origin) who owned some property, knew their neighbours, shared common problems and often a common occupation with them, and were used to collective action in guilds and confraternities. The whole organization of industry in Paris fostered these conditions. Work places were small. Reveillon's factory, with its 350 workers, was quite exceptional. There were only a handful of establishments in the capital whose labour force ran into three figures, and none employed more than a thousand. The average workshop employed between 16 and 17 artisans, and in the whole of Paris only 473 establishments had a work force of over 30. The workshops of certain trades tended to concentrate in certain districts, and their personnel were also bound together, in almost all trades requiring any skill, by guilds. Guild membership was estimated by the police lieutenant at over 100,000, or a third of the male population, in 1776.[6] It is true that guild organization was not popular among many, perhaps most, of those involved in it. Journeymen and apprentices resented the power vested by guilds in masters, and the difficulty of rising to masterships themselves. There were jubilant street demonstrations when Turgot abolished guilds in 1776, and a wave of industrial unrest when they were re-established later in the same year. The forces of public order, for their part, regarded the guilds as an essential instrument for controlling the workers, bitterly opposed Turgot's abolition, and tried to tighten guild organization when it was restored. The fact remains that guilds gave their members important economic privileges, accustomed them to working together for common objectives, and bred habits of organization—if only by provoking their subordinate members to combine against them.

The police authorities and the courts of law drew no distinction between industrial action and breaches of public order. They regarded the working population as potential savages whose bestial and anarchic instincts could only be contained by constant vigilance and laws of draconian severity. Throughout the century there was a steady trend towards ever closer supervision of the world of work.[7] The disturbances of 1776 only confirmed the belief that authority must never relax its grip, and in 1781 a refinement was introduced in the form of the *livret*, a work record which every artisan was obliged to carry and have endorsed by his successive employers. He would, of course, receive no endorsement, and so would theoretically be unable to find another employer if he proved insubordinate. And yet insubordination, too, appears to have been on the increase during the later decades of the century. Strikes, though mostly short ones, were frequent, and although the authorities invariably supported the employers with a whole range of legal threats, occasionally strikes succeeded.[8] Their aims were familiar enough—reductions in the length of the working day, resistance to wage cuts, protection of a trade against new forms of competition, and just occasionally support for wage rises. No doubt the economic uncertainties of these years, and the long-term decline in the purchasing power of wages, were the main causes of this growing unrest. They were certainly far more important than the 'subversive literature' blamed by literary observers with a built-in but excessive belief in the power of the written word. What is clear, however, is that the first instinct of Parisian workers faced with economic hardship was not to put pressure on their employers. When the price of bread went up they did not seek higher wages in order to pay for it; they tried to force the government to bring the price down.

The authorities knew that the true foundation of public order in the capital was a guaranteed regular bread supply at prices which ordinary people could afford. They regarded this as so important that they would not hesitate to see other areas go short in order to keep Paris supplied. For thirty miles around the city its needs had absolute priority, and privileged private companies were employed to buy supplies further afield. At every stage, from its entry into the city right through to its appearance in the shops as loaves of bread, the trade in grain was carefully supervised by the city authorities; and the price of bread was officially fixed every day. This price bore no direct relation to that of grain. The authorities did not hesitate to hold it down by means of subsidies if this appeared to be in the interests of public order. But sooner or later if harvests were bad the price of bread had to rise, and it was then, as we have already seen, that the people of Paris grew restless, and disturbances occurred. The most dangerous days were market days, when large crowds of consumers gathered close to where bread or grain was being sold. It was then that rioters, often led by women, intimidated grain merchants and bakers, and tried to force them to sell their wares at a price below the official one. In times of high bread prices, too, popular resentment turned against the entry taxes (*octrois*) which pushed up the price of all goods, including foodstuffs, entering Paris. As the city expanded over the century these dues had become increasingly difficult to collect and easy for smugglers to evade on a massive scale. The complaints of the Farmers General of taxes, who managed the system, grew insistent, and in 1785 Calonne authorized them to build a ten-foot wall around the city punctuated by purpose-built gates (*barrières*) for the collection of dues. It was still going up in 1789 and a subject of intense popular resentment.

Yet although the importance of keeping Paris calm was never out of the government's mind, few people believed that it would ever lose the upper hand. 'Dangerous rioting', wrote the journalist, Mercier, in his sketches of metropolitan life, 'has become a moral impossibility in Paris. The eternally watchful police, two regiments of Guards, Swiss and French, in barracks near at hand, the King's bodyguard, the fortresses which ring the capital round, together with countless individuals whose interest links them to Versailles; all these factors make the chance of any serious rising altogether remote . . . . Any attempt at sedition here would be nipped in the bud; Paris need never fear an outbreak such as Lord George Gordon recently led in London.'[9] Yet the city's regular police force, organized into several independent units which seldom worked closely together, numbered less than 2,000,[10] and the Swiss guards and the royal bodyguard were at Versailles, a three-hour ride or five-hour march away. The ultimate safeguard against disorder, therefore, was the 3,600-strong regiment of the French Guards permanently billeted in the capital. These troops were not called out often between 1775 and 1787, and their relations with the Parisians were good. They were not, like most other units, frequently moved around from one posting to another, and they did not live in barracks. Many were married, and many earned extra money pursuing civilian trades in their abundant spare time. None of this made them unreliable. They helped to contain anti-government riots throughout 1787 and 1788, and it was they who played the main role in putting down the Reveillon riots as late as the end of April 1789. But, living in such close daily contact with those they were supposed to contain, they were subject to the same hopes, fears, and economic pressures, and demonstrators soon grew adept at appealing to them as fellow citizens with common interests.

The people of Paris took little part in precipitating the collapse of the government in August 1788. They gave fairly consistent support to the parlement in its opposition to Brienne, but support for the parlement was a traditional popular reaction in times of political conflict. It enabled them to let off steam with the excuse of supporting the guardians of the law and the liberties of Frenchmen against the encroachments of authority. Besides, the parlement was well known for its hostility to decontrolling the grain trade.[11] But riots in support of the parlement in August and September 1787 had no effect on the government's policy, and eventually the French Guards cleared the streets. A year later, several weeks of celebrations greeted Brienne's fall and the restoration of Necker, and this time a number of demonstrators were killed before order was restored. Apparently Necker was alarmed at the scale of the demonstrations, and they may have played a part in his decision to suspend all reforms and advance the date of the Estates to 1 January.[12] Certainly the repeated tumults of these years had a cumulative importance. Not since the early 1770s had they been so frequent, and now they came at a time when public excitement had already been whipped up by a prolonged political crisis and a torrent of publicity. By the spring of 1789, especially as bread prices rose to levels not seen for almost twenty years, there was a widespread expectation that sooner or later the people would take to the streets; and it had begun to occur to some of the more extravagant leaders of the 'patriot' party that it might not be altogether unwelcome if they did.

These leaders were not to be found in the committee of Thirty, who had started the agitation for doubling the third estate, or even among the third estate deputies convening at Versailles in May 1789. By then the forcing-house of political radicalism was in the cafés of the Palais Royal. The Palais Royal was the Paris mansion of the duc d'Orléans, and in 1780 its gardens and galleries were thrown open to the general public. They rapidly became the social centre of Paris. Their cafés and bookshops made them a clearing house of publicity, information, and rumour, and because they remained technically Orléans's private property, the police were not allowed there. Consequently, the Palais Royal was also a happy hunting ground for fugitives, pickpockets, and prostitutes. And although it was not often frequented by the ordinary working people of Paris, being far from the eastern districts in which most of them lived, it was a popular entertainment centre on rest days, and there was a bread market there. Above all, in the spring of 1789 it was the place to go to learn the latest news and discuss the latest opinions. 'The coffee-houses in the Palais-Royal', wrote Young on 9 June, 'are not only crowded within, but other expectant crowds are at the doors and windows, listening *à gorge déployée* to certain orators, who from chairs or tables harangue each his little audience. The eagerness with which they are heard, and the thunder of applause they receive for every sentiment of more than common hardiness or violence against the present government, cannot easily be imagined. I am all amazement at the ministry permitting such nests and hotbeds of sedition and revolt . . . it seems little short of madness. . . .'[13]

The Reveillon riots broke out far from the Palais Royal, in

the heart of the manufacturing faubourg Saint-Antoine. Apart from the fact that they were precipitated by a remark made in an electoral assembly, and perhaps owed something to frustration at the news of a week's postponement in the meeting of the Estates, they were curiously divorced from the political excitement of the time. Basically they were an old-fashioned bread riot, and unlike their predecessors in September 1788, or their successors in June and July 1789, those participating had no political objectives. All the same, the riots did have important political repercussions. They threw the shadow of popular disorder over the elections for the Paris deputies to the Estates, which were not completed until the end of May. They reminded the members of the Estates deadlocked for almost two months, that Paris was a powder keg close enough to blow them all away if it exploded again. Above all, perhaps, the government's grudging and ungenerous attitude towards the French Guards, who had brought the disturbances under control, began a disintegration in that regiment's discipline and morale that was to prove disastrous a few weeks later.

While the rest of France waited tensely for the Estates-General to do something during the six weeks after they first convened, the still-uncompleted elections kept up the level of political activity in the capital. And, despite frantic efforts by Necker to bring in extra grain supplies from abroad, the price of bread continued to rise. It was to reach its highest point of the year on 14 July, when a loaf cost 16 *sous*, a level not seen since 1770. In these circumstances it seemed only prudent to reinforce the police, and Besenval, the military commander of the Paris region, brought 4,000 extra troops to within marching distance of the city between mid-April and the end of June. These measures were quite distinct from the troop movements ordered by anti-Neckerite ministers in subsequent weeks, but Paris did not draw the distinction. In the atmosphere of frustration, uncertainty, and growing suspicion that characterized these weeks of inaction at Versailles, troop movements looked like evidence of a plot to dissolve the Estates by force, and the Royal Session announced for 23 June looked like the moment when this attempt should be made. The absence of Necker from the session suggested that he had been dismissed, and once this news reached the Palais Royal there was uproar. Among those carried away with indignation were members of the French Guards, and on the 24th two companies declared that they would not act to prevent popular demonstrations. On the 27th, five more companies simply threw down their arms and made their way to the Palais Royal, where they were greeted as heroes. 'It may well be conceived', wrote the appalled British ambassador, 'the effect this had upon the populace who now became quite ungovernable at Versailles, as well as at Paris.'[14] It was fear that the troops would decline to protect him that induced Louis XVI to command the three orders to unite on 27 June. The ringleaders of the mutiny were promptly arrested, but when the news reached the Palais Royal a huge crowd marched to where they were imprisoned and forced their release. Public collections were organized on their behalf, as well as a movement to petition the king to pardon them. The third estate electors of Paris, who had continued to meet after deputies had been named, and even the National Assembly itself, sent in petitions. Clearly the Palais Royal had become a sort of headquarters from which 'patriotic' action in Paris and Versailles was being planned and coordinated.

For every mutinous soldier, however, there were hundreds whose discipline remained firm,[15] and during the first week of July they were arriving thick and fast. On 6 July they cleared the Palais Royal, though without firing. On the 8th they were called out to enforce payment of dues at various customs barriers, where crowds had gathered in the hope of forcing the admittance of cheaper foodstuffs. The same day the frightened Assembly petitioned the king to withdraw all troops thirty miles from the capital, and was not reassured by his reply that they were necessary to maintain public order. And so they were, if Louis and his aristocratic advisers were to have any hope of succeeding in their counter-revolutionary designs. What precisely these were is not entirely clear, since they were never put into effect. But the essential preliminary step was the dismissal of Necker, which was decided on 8 July. He was formally notified, and told to leave the country at once, on the 11th, and on the advice of the queen and the king's brother the count d'Artois a new, conservative ministry was constructed under Calonne's old rival, Breteuil. The intention to enforce policy with troops was shown by the appointment of the veteran marshal de Broglie, whom Louis had summoned to Versailles two weeks previously as 'a man he could trust', as secretary for war.

Despite his dismal and vacillating performance since the opening of the Estates, Necker had now reached the height of his popularity. Ever since 23 June, he had been the hero of the Palais Royal, and those who dared to criticize him were beaten up. He was regarded as the people's protector in the government, the guarantor of bread supplies, the third estate's champion; and the king must have known that to dismiss him was to throw down a challenge. At first the *coup* looked like succeeding for all that. When the news of Necker's dismissal broke at the Palais Royal on the 12th–a Sunday, incidentally, when nobody was at work–a predictable demonstration began. Theatres were closed as a sign of mourning, and five or six thousand people marched to the Tuileries gardens behind a bust of Necker. There they were set upon by a German cavalry regiment which only withdrew to await reinforcements when the crowd fled to higher ground. By the time the reinforcements arrived it was dark and, uncertain of what they faced, their commanders decided not to continue with the operation. But blood had been shed, and it seemed obvious that the forces ringing Paris were about to go into action. The capital was surrounded. The scarcity of bread aroused suspicions that the people were already being deliberately starved. If the people of Paris did not defend themselves now, nobody knew what might happen.

Thus began the insurrection which culminated in the taking of the Bastille on 14 July. Primarily it was a search for arms. Throughout the 13th, every place in the city where arms where known, or even suspected, to be stored was ransacked. The Bastille was merely the last, and the most formidable, depot that the rebels reached, and it was only taken by storm because its terrified garrison made a half-hearted attempt at resistance. Only later did it come to seem especially significant that an apparently impregnable royal fortress, and a sinister state prison where victims of tyranny had often languished, should have fallen to popular assault. It is, indeed, unlikely that it would have fallen had the efforts of the besiegers not been co-ordinated by mutinous members of the French Guards, professional soldiers who knew what they were doing. But the insurrection was also a food riot. During the night of 12/13 July, 40 out of the 54 gates in the new customs wall were attacked and burned, and at certain points the wall itself was demolished. A later investigation of those known to have taken part in these attacks showed them to be people who were hard hit by the higher prices of wine, foodstuffs, and firewood that resulted from the *octrois*: they were shopkeepers, petty tradesmen, and wage-earners.[16] The abbey of Saint-Lazare was also looted after rumours spread that grain was stored there. Now they had armed themselves, it was too much to expect the people of Paris not to take direct action on their most pressing problem, that of feeding themselves.

The looting and destruction were spectacular, but they were far from an outburst of indiscriminate savagery. 'Nothing', wrote the extremely aristocratic British ambassador on 16 July, 'could exceed the regularity and good order with which all this extraordinary business has been conducted: of this I have myself been a witness upon several occasions during the last three days as I have passed through the streets, nor had I at any moment reason to be alarmed for my personal safety.'[17] Yet substantial, propertied Parisians *were* alarmed. When food riots broke out they always were. The third estate electors had shown their fear when on 11 July they began to discuss setting up a citizen militia to keep order, as had already been done in Marseilles and several other towns. On the 12th they reconvened and decided to convoke meetings of citizens in all the sixty electoral districts for the next morning. It was done by ringing the *tocsin* (alarm bell) from the churches and firing cannon, which only helped to perpetuate the atmosphere of crisis and panic. Later on the 13th, electors meeting at the town hall decided formally to set up a militia under the orders of a committee elected by themselves. The task of finding arms for this new body did much to maintain the momentum of the insurrection throughout the next day. The important thing, however, is that this self-appointed committee was obeyed. It had in fact taken over power in Paris; and although its main concern was to control popular violence, it was popular support that gave it authority during these first turbulent days.

Nor was there any power that could reverse this takeover. The withdrawal of the troops on the night of 12 July proved fatal. There is no evidence that most of them would not have performed their police duties with complete obedience if they had been sent back into the city; but their officers no longer had confidence in their reliability. Strenuous attempts were certainly being made to subvert them by appeals to their 'patriotism', and by 12 July other regiments besides the French Guards were experiencing desertions. All this scared the officers, and they began to advise their superiors that the men were not to be depended on. On 16 July, Broglie advised Louis XVI that he could not trust his army, without ever seriously putting this suspicion to the test. In taking this advice, Louis was admitting that he had no means of enforcing his will against Paris, or indeed against anybody. All hope of arresting the course of the Revolution was now gone. Louis went to the Assembly and announced that the troops would pull back. Necker was recalled. On 17 July, the king went to Paris and confirmed all that had been done.

Events did not stop short here, but they would have done if the policies of Artois and the queen had been carried through. In this sense the Revolution had survived the crisis of its birth, and was now launched. The National Assembly was saved, and representatives of the propertied nation had taken control of affairs not only there, but in the capital city too. In subsequent weeks the example was to be followed all over France, as self-appointed local notables deposed the old municipal authorities without meeting any resistance, and set up units of the National Guard (as the citizens' militias were now known) to back up their authority and maintain public order. No doubt they would have preferred this transfer of power to have taken place in a more orderly and regular way.

They would certainly have preferred to take control without the need for arming the fickle and unpredictable populace. But without the intervention of the hungry people of Paris, the bid for power that had begun on 10 June might well have been defeated, and this was a fact that the deputies had to accept. The Parisians did not forget it either. They knew they had saved the Revolution, and they were proud of it. What they had done once they could, and did, do again. For good or ill, the French Revolution was to be characterized by regular popular intervention.

Meanwhile the work of national regeneration could really begin. On 7 July, the deputies renamed themselves the National *Constituent* Assembly, to emphasize that their essential task was to give the country a constitution. On the 8th, they abrogated all binding mandates, so as to leave every deputy free to follow his own convictions in framing the constitution. On 4 August, after prolonged debate, it was agreed to begin the constitution with a Declaration of the Rights of Man and the Citizen, a manifesto of what the Revolution stood for. Before twenty-four hours had elapsed, however, it stood for far more than many of them had ever expected; for that night 'feudalism', and much else, too, was abolished. This was largely the achievement of the one group who until now had had no say in what happened—the peasantry.

12. The Peasantry

Many analyses of the origins of the Revolution begin with the peasantry and its problems and grievances. The logic of this no doubt lies in the fact that over 80 per cent of Frenchmen were peasants. It certainly cannot lie in any role that the peasantry might have played in precipitating the old regime's final crisis, since they played no role. Right down until the spring of 1789, the peasants were completely passive observers of what was happening, and those involved in the political struggles of the pre-revolutionary years gave little thought to them. What made the peasants important was the failure of the 1788 harvest, which endangered public order in the countryside quite as much as in the towns; and the drawing up of the *cahiers* in the spring of 1789, which raised expectations, as a peasant woman told Arthur Young, that 'something was to be done by some great folks for such poor ones'.[1] When these expectations were not immediately fulfilled, they took the law into their own hands, and the National Assembly was forced to appease them in ways which most of its members would have preferred to avoid.

'Peasant' is a vague term, harder to define than noble but not by any means as difficult as bourgeois. Peasants were country dwellers; but within that broad definition, as with other social categories, there was an enormous range of wealth, status, and outlook. The most obvious way of subdividing the peasantry is in terms of the land to which they had access. This is not the same as the land they owned, for all the evidence suggests that enjoyment of land rather than ownership was the crucial thing. Some four million peasants actually owned perhaps a quarter of the land of France, a proportion rather less than that of 1700; but almost all the other land was worked and exploited by peasants under leases of one sort or another. Everything depended on the size of the units in which it was owned or leased. At the apex of the peasantry stood a small group of large-scale farmers, not always easily distinguishable from bourgeois. They tended to exploit large concentrations of land, or take the leases of whole farms or even estates from wealthy absentees, subletting to smaller men in poorer areas, or exploiting on a big scale for the market in the opulent grain-growing areas of the north and north-east. This rural élite numbered scarcely more than 600,000, but everywhere its members dominated local life, especially if noble or bourgeois landowners were absentees. More numerous, but still accounting for only a minority of the peasantry, were those with enough land for self-sufficiency, or even a modest saleable surplus in a good year. They were often known as *laboureurs*, and were viewed with envy and respect by the rest of the rural community, for whom self-sufficiency was a distant and increasingly unattainable dream. Big farmers and *laboureurs* did very well over much of the eighteenth century. Surpluses sold profitably in the boom conditions down to the 1770s, and they were cushioned against the recession of the subsequent two decades by their ability to live off their own produce, accumulate stocks, and, if they sublet, to increase rents.

But the vast majority of the peasantry fell into neither of these groups. Most of them enjoyed some land, but not enough for self-sufficiency. They were therefore compelled to rely upon other activities in order to survive. They earned money as wage labourers, plied rural crafts, wove cloth for urban *entrepreneurs*, migrated seasonally, borrowed from money-lenders, and if the worst came to the worst, they took to begging. Theirs was 'an economy of makeshifts',[2] providing a bare living in good times, but one inadequate to sustain them in bad; depending on the labour of the whole family and likely to collapse if any of them were disabled or if there were too many unproductive children. From here the slide was almost inexorable into the lowest category of peasants, the landless, with nothing to depend on but the strength of their arms and the quickness of their wits. This group was the heart of the rural poor, a group varying in size according to the fluctuations in the economy but always running into several millions. Among them were numbered most of the beggars, vagrants, and petty criminals in the countryside as well as the bulk of the casual labour force in country and town alike. Without land they were without roots, and so were difficult to enumerate and almost impossible to police. More sedentary peasants were afraid and suspicious of them.

These broad categories conceal a wealth of regional diversity that would take volumes to describe. But every region, and every category of peasant, was affected by certain basic trends over the eighteenth century. Fundamental among these was the rise in population.[3] Most of it took place in the countryside. Urban population grew through immigration, itself suggestive of rural over-population. In everyday terms population growth meant larger families and more mouths to feed, and this put increasing pressure on the economy of peasant families. More and more people were now seeking land; even those who had hitherto enjoyed enough now needed more. And yet the law of succession, which was more or less divisive everywhere, meant that holdings became increasingly subdivided and accordingly less rather than more adequate for the needs of those working them. In these circumstances it was easy for the rich to snap up uneconomic lots, thus diminishing further the stock of land available. What remained available, meanwhile, became the object of increasingly desperate competition. Rents were driven up, and sharp clashes occurred over the use of common lands and common rights. The government, under physiocratic influence, believed common rights to be harmful to agriculture, and in the 1760s it issued edicts allowing large owners to challenge them. It also encouraged the enclosure of common lands, which poorer peasants tended to welcome, but which aroused mixed feelings among the rich and so made limited progress.[4] By far the most important result of population growth, however, was an enormous expansion in the ranks of the landless, of migrant workers, of beggars, and of vagabonds. Times of sudden economic crisis, such as the spring of 1789, swelled their numbers yet further.

Abundant labour was cheap labour; and accordingly wages rose slowly, at only one-third the rate of prices. Every peasant who was not minimally self-sufficient suffered from this trend, and that meant the vast majority who had land, as well as the landless. And in the two decades before 1789, the standard supplements relied upon by those without enough land to guarantee independence–wine production and weaving–underwent unprecedented fluctuations which left them even less well prepared than usual to face difficulties. Even migration to the towns held less promise, since the economic crisis diminished urban employment prospects. And meanwhile the cost of living rose inexorably. Peasants who were not self-sufficient had to buy a proportion of their needs and so were hit by the 62 per cent price increase over the century. They also suffered from the rise in taxes that took place between 1749 and the 1780s. Even if inflation made the real impact of this rise less serious than its nominal one,[5] the peasants bore proportionately more of the burden than any other group. They alone enjoyed no sort of exemption; they alone bore the full burden of the *gabelle* (state salt monopoly) or the draw for the militia, which arbitrarily took able-bodied young men away from the land. Meanwhile the advantages enjoyed by the privileged were in flagrant contrast with these conditions. In villages near towns, the burden of the *taille*, for example, was often increased by rich nobles or bourgeois enjoying exemption who bought lands which thereby acquired exemption, too, leaving the poorer inhabitants to redistribute a heavier burden among themselves. In the 1770s and 80s, ministers and intendants often congratulated themselves on their humanity in attempting to commute the *corvée*, forced labour service on the public roads, into a money payment. But many peasants preferred to continue labouring rather than pay yet another tax. Then there were tithes, which all those with land had to pay, for the upkeep of the parish clergy. They were nominally a tenth of all fruits, although in practice they usually amounted to rather less, but they were payable in kind, and therefore were only inflation-proof to the beneficiary. Yet the chief beneficiary was often not the incumbent at all, but a monastic or lay impropriator who paid the priest only a *portion congrue*. Finally there was the weight of seignorial dues, the whole system of what contemporaries called 'feudalism'.

Feudalism in 1789 was a relic of the medieval system of land tenure. But it was far from an empty relic. All land subject to feudal tenure—and by 1789 only a tiny proportion was not—had a lord, and the lord exercised certain rights over it. But long before the eighteenth century lordship and ownership had ceased to be synonymous. It was rare to find a lord who did not own any of the land over which he exercised rights; but it was equally rare to find a lord who exercised rights only over his own land. Most landowners, therefore, found their property rights circumscribed by obligations to a lord, or several lords. And whereas by the eighteenth century the pattern of property had become so complicated that large proprietors often found themselves both lords and vassals at the same time, peasants never did. The trappings of lordship were completely beyond their pockets, even if the idea of peasant lords had been socially conceivable, which it was not. 'Feudal' obligations were one more burden that the peasants bore the full weight of alone.

Feudal rights fell into three broad categories.[6] Some were purely honorific, such as the right to precedence on public occasions, a seignorial pew or a hatchment for arms in the parish church, and the right to erect dovecotes, or weather-vanes on the manor-house. Such things were far from trivial, for they were the outward symbols of power. The zeal with which peasants attacked displays of arms and weather-vanes in the Revolution shows that they, as well as the owners of these vanities, took them seriously. Secondly, there were jurisdictional rights. Most local courts of first instance in the countryside were lords' courts, with privately appointed judges and officials and considerable coercive powers over vassals. What made them particularly odious to the peasantry was that they made lords judges in their own cause, since all defaults on feudal obligations were tried there. Thirdly, and most important, there were so-called 'useful rights'. These included hunting rights, shooting rights, and monopolies of village milling, baking, and wine pressing (*banalités*). Above all, they included the right to levy dues in cash or kind. The level and nature of these dues were infinitely variable, depending on the custom of each individual lordship. Their over-all general weight also varied hugely from region to region. In Burgundy and Brittany they tended to weigh very heavily, yet in areas on the very borders of Brittany, for instance, they were so light as to be barely noticeable. In the Auvergne, they have been estimated at just under 10 per cent of the peasants' gross product, in the south-west at nearer 11 per cent, and in the Toulouse region at just over 15 per cent.[7] All these figures conceal staggering variations from one community to another, and generations of historians have been baffled by the difficulty of coming to any reliable general conclusions. One thing contemporary peasants were sure of was that feudal dues were inseparable from the nobility. Originally all lords had been nobles by definition, and in the eighteenth century probably most still were. Lordship was a dignity that no noble felt authentic without; and besides, dues could be profitable. In the Toulouse region they brought in an average of almost 19 per cent of the gross revenues of forty-eight lordships in 1750.[8] Such returns made 'feudalism' a good investment, and accordingly the number of lords who were not nobles at all grew steadily. They included monasteries and other ecclesiastical corporations as well as many bourgeois who saw the double value of feudal rights as profitable property and a symbol of social prestige.

The feudalism of the eighteenth century, therefore, bore little resemblance to its medieval progenitor, but in contemporary terms it was very much a going concern. And over the century its burden appears to have grown heavier. For long this was attributed to a 'feudal reaction' on the part of noble landlords; but this may be an illusion.[9] Its reality has yet to be demonstrated.[10] Yet surveying techniques and archival ingenuity undoubtedly improved, which meant that when the terriers which recorded feudal obligations were renewed, as they had to be every generation to be of use, it was done with increasing precision. *Feudistes,* agents who lived by revising terriers, were more professional and therefore more formidable than they had been in the previous century; and more lords may have been farming out the collection of their dues to middle men whose only interest was in making a profit during the period of their lease.[11] Inflation certainly increased the burden of dues payable in kind; and finally, in 1786, a new tariff of duties was introduced for the work of terrier revision. The vassal, not the lord, paid these charges;

and the *cahiers* show that many peasants felt they had suffered from recent revision of terriers.

Even so, the number of peasant *cahiers* which condemned feudalism as a whole was surprisingly small.[12] Such an idea was beyond the intellectual grasp of illiterate or semi-literate peasants, and certainly far beyond what most of them could have dreamed was possible. Most criticism concentrated on specific manifestations of feudalism, such as seignorial courts, or *banalités*, or regional peculiarities such as the serfdom common in Franche Comté. Only where model *cahiers* were circulated by town-based radicals were peasant minds opened to the possibility of comprehensive changes; and it is a curious fact that the most persistent condemnation of feudal practices came not from rural *cahiers* at all, but from urban ones. No doubt this reflects the fact that feudalism had come under intellectual attack in the course of the century, notably from physiocrats who saw rights and dues as harmful burdens on agriculture.[13] The elections of 1789 also brought feudal categories to public notice in that one qualification for voting in the assemblies of the nobility was holding a fief. Any movement questioning the prerogatives of the nobility, indeed, could not fail to provoke reflection on a system of rights with which it was so closely identified. Even so, there is no strong evidence of a groundswell of hostility to feudalism as a whole among the literate classes. The intellectual attack seems to have lost momentum after the fall of Turgot, and too many bourgeois had invested in feudal rights to begrudge them to the nobility. They were, after all, property, and it was generally agreed that they neither could nor ought to be abolished without compensation. Even among the peasantry who bore most of the burden, feudal rights were scarcely questioned spontaneously. It took the crisis of the spring of 1789 to push them into this.

Hunger, hope, and fear were the main ingredients of the rural crisis of 1789.[14] The economic crisis was responsible for the hunger. Because most peasants were not self-sufficient, they had to buy their bread and flour just like town dwellers, and so they were equally badly hit by scarcity and high prices. As the backbone, too, of France's industrial labour force, they suffered from the fall in demand for textiles, and were less able to afford the rising price of what bread there was. So from January onwards there were disturbances up and down the country, often beginning on market days. Grain convoys were attacked, suspected hoarders intimidated and their premises ransacked, millers and bakers were harassed, and local authorities threatened. Yet thus far the course of events was predictable enough. What transformed the normal consequences of a poor harvest into something more serious was the electoral campaign. The elections, and the promise of the Estates-General, raised vague but powerful hopes that better times were just around the corner. Every male taxpayer over twenty-five could cast a vote and state his grievances in a *cahier*. The king appeared to be taking a genuine interest in his subjects' problems, and seemed to be promising relief. A 'great hope' suffused the countryside that the miseries of the last few years would soon be over, and that the problems of poor peasants, old and new, would soon be solved.[15] In many areas, indeed, it was obviously believed that to state a grievance in a *cahier* was automatically to have it redressed. 'All the peasants around here', wrote a Breton subdelegate on 4 July, 'are preparing to refuse their quota of sheaves to the tithe collectors and say quite openly that there will be no collection without bloodshed on the senseless grounds that as the request for the abolition of these tithes was included in the *cahiers*, . . . such an abolition has now come into effect.'[16]

In the seventeenth century, when peasants rebelled, their main grievance was usually the increasing weight of taxation. This grievance was still there in 1789. Like the people of Paris who burned down the customs posts on 12 and 13 July, the peasants believed that indirect taxes made necessities unnecessarily expensive. They also believed that direct taxes made them less able to bear the soaring price of food. Accordingly, the spring of 1789 witnessed the beginning of a nation-wide 'tax revolt' against all forms of government imposition.[17] People simply stopped paying taxes, direct or indirect, and hapless collectors, without the support of any firm central authority, were unable to force them to do so. But this time it was not only taxes that they refused to pay. In contrast with their forefathers in the previous century, the peasantry of 1789 also turned against tithes, and above all against feudal dues.

Refusal to pay dues was reported from Dauphiné as early as 13 February. In March, peasants hunting in the Prince de Condé's woods near Chantilly killed two gamekeepers who tried to stop them, while in Provence a noble who tried to prevent peasants from ransacking his château was murdered. In April, vassals in the Alpine region began to take back by force the grain which they had paid over as dues in the previous year. Throughout Provence there were attacks on bishops' palaces and monastic barns, invariably accompanied by declarations that no tithes would be paid in future. Nor did the meeting of the Estates stem the flow of incidents. Initially, at least, it only raised expectations still further and encouraged peasants to anticipate what they were convinced would be the deputies' decision to emancipate them from their burdens. Subsequently, as nothing was achieved at Versailles, the peasantry grew impatient and suspicious, and rumours began to reach the provinces from the capital telling of aristocratic plots to obstruct the work of national regeneration. The chosen instruments of aristocratic revenge—so ran the rumours—were to be 'brigands'; and the countryside was certainly overrun that spring with vagrants, vagabonds, and migrants in search of work. These categories were all very different, but the peasantry were in no mood to draw fine distinctions. They were all unknown strangers on whom suspicion easily fell. Anxieties were only increased as the grain began to ripen in June and July. The harvest promised to be abundant, bringing twelve lean months to an end; but if the 'aristocrats' were plotting to reassert their authority, one obvious means was presumed to be by starving the populace into submission by cutting down unripe corn. Suspicion bred credulity; and as the stalemate continued at Versailles, and the complaints of indignant third estate deputies grew ever more strident, the conditions ripened for a wave of rural panic.

The political crisis that lasted from the Royal Session of 23 June until 16 July brought all these tensions out. The aristocratic plot seemed about to take effect. Thousands believed this in Paris and Versailles, so it is hardly surprising that the provinces should accept it uncritically, too. The news of the Parisian uprising on 12 July came as a final spark to a fire that was already smouldering, and over the next three weeks it seemed as if order had practically collapsed in the countryside. The incidents of the spring against taxes, tithes, and feudal dues now became a general movement. In particular, there was a wave of attacks on seignorial castles and manor houses.[18] Hundreds were ransacked and many were set on fire. Yet it was not a movement of indiscriminate destruction. It is true that insurgents often searched for hoards of grain, but what they really wanted were terriers and other written records of feudal obligation. When found, these documents were promptly taken out and destroyed. When not, the château was burned down so as to destroy them in any case. Sometimes, too, lords were compelled to sign declarations that they would not attempt to reimpose dues. Sometimes they were forced to take down weather-vanes or even slaughter their pigeons. But always the objective was clear—not only to end feudal dues but also to make it impossible to reintroduce them even if the old order were restored. The determination of the peasantry was only reinforced as the news filtered through of the king's capitulation to the Parisian revolt; it was taken as a sign that he approved, indeed had authorized, the use of direct action against the pretensions of aristocrats. Certainly this conviction lent legitimacy to acts of violence where they occurred. Yet on the whole it was a remarkably orderly business, the insurgents' objectives were clear and limited, and hardly any blood was shed. Only where lords or their agents attempted to resist was there violence, and even then it was against property rather than persons.

But this was no solace to property owners. Large and small alike felt at risk, and peasants who were not involved in attacks on their lords were no less panic-stricken than the lords themselves. In fact, the revolts took place against a background of popular hysteria known to historians as the Great Fear.[19] One of its mainsprings was the rumour that château burning was the work not of fellow peasants at all, but of the universally dreaded brigands. It was taken as sure evidence that they were coming to murder, pillage, and destroy the ripening crops. The fear compounded the chaos in the countryside, and in Dauphiné château burning began only after the peasants had been armed to fight brigands who never came, because they never existed. And meanwhile at Versailles, the deputies were increasingly prey to a great fear of their own—that law and order were breaking down completely and irrecoverably. Wild charges and counter-charges were exchanged over who was responsible for the destruction. Anguished nobles blamed bourgeois plots; third estate leaders blamed aristocratic ones. Nobody appeared to believe that the peasantry were acting for themselves. This apparent rural anarchy, however, seemed to herald the catastrophe which had haunted deputies of every order since the Estates had convened—a situation of such chaos that all forms of property were at risk.

It was to prevent this that feudalism was abolished on 4 August. As already noted,[20] there was no mandate for such abolition in the *cahiers*. Only specific manifestations of feudalism, such as serfdom, seignorial courts, hunting rights, or *banalités* were widely condemned. This was enough to scare the nobility, however, and one reason why they were reluctant to agree to vote by head was the fear that the third's natural majority would bulldoze through a blanket abolition. For the same reason the royal programme enunciated on 23 June specifically guaranteed that feudal and seignorial property, useful rights, and the honorific privileges of the first two orders would not be subject to discussion in common. Yet the third estate showed little interest in these questions. There seems to have been a general assumption that sooner or later the more glaring anomalies of feudalism such as serfdom and hunting monopolies, would go, but nobody mentioned dues. The deputies of the third had too strong a sense of property to think of abolishing them. But this same sense made them appalled by the reports of destruction that were pouring in from the countryside. Their first instinct was to try to expedite the organization of the National Guard, but it was soon clear that this could not be done quickly enough. Exhortations to patience were also discussed at length during the last week in July, and it was during these discussions that the idea of abolishing feudal dues first came up. It proved very contentious. Some, and not all of them nobles, asserted that these property rights could not simply be abolished. Others argued that this alone would mollify the peasants, since it was all they cared about. Eventually a group of radical deputies decided formally to move that dues should be abolished subject to compensation, and labour services abolished outright. To increase the likelihood of success, it was agreed that a sympathetic nobleman should launch the debate late one evening when the Assembly was poorly attended.

Such were the origins of the session of the night of 4 August.[21] But almost as soon as it began the carefully laid plans went awry. The measured proposals with which the sitting began soon snowballed into a series of more grandiloquent gestures. It was agreed to abolish all feudal obligations outright, then noble privileges were offered up, then provincial and municipal ones. Tithes were abandoned, too, as were other clerical revenues. Venality of office followed. The deputies were swept along by a wave of altruistic enthusiasm. 'Everything', wrote a moderate, sceptical nobleman, 'prescribed to us the conduct we must pursue; there was but one general movement. The Clergy, the Nobility, rose up and adopted all the motions proposed . . . it was the moment of patriotic drunkenness.' Nor was he immune to these feelings himself. 'This union of interests, this unity of the whole of France to the same goal (the common advantage of all), that twelve centuries, the same religion, the habit of the same ways, had been unable to bring about . . . found itself formed all at once, sanctioned forever.'[22] The fact that calculating men used the occasion to carry changes they had not previously thought possible, or to decree away the property of their enemies, does not detract from the 'magic' quality of the occasion. Cooler ploys could not have succeeded had the deputies been less excited. Yet it all contributed to the general destruction of rights and privileges. When the deputies awoke on the morning of 5 August, they found that they had abolished most of the central social institutions of France. At a blow, these things had become the *Ancien Régime*. It was to be the work of the Revolution to decide how to replace them.

13. Conclusion: The New Regime and Its Principles

By August 1789, the *Ancien Régime* had gone beyond recall. The French Revolution was irreversibly launched. It is true that Louis XVI and his courtiers remained basically unreconciled to what had taken place, and that late in September they began to toy once more with the idea of recovering control by military force. The people of Paris did not feel that the Revolution was safely established until they had brought both king and National Assembly back to the capital in the October Days.[1] But even if this had not happened, it seems unlikely that anything short of prolonged civil war could have turned the Revolution back. By then every locality was in the hands of notables committed to the new order and disposing of a National Guard that was better organized with every day that passed. And people spoke of the Revolution as an accomplished fact. What the Parisians thought they were saving in the October Days, and what the aristocratic *émigrés* who began to trickle out of France after 14 July felt unable to live with, were the achievements of that summer. The study of the Revolution's origins, therefore, properly ends then. Everything that happened later, even counter-revolutionary plots, is part of the Revolution's own development. After all, the very idea of counter-revolution implies the prior existence of a Revolution to counter.

France, then, was now governed, as it was to be for the next decade, by people who thought of themselves as revolutionaries and acknowledged the inspiration of certain guiding principles established in 1789. What were these principles, and why were they adopted? Essentially, they were enshrined in two documents: the decree of 11 August abolishing feudalism, and the Declaration of the Rights of Man and the Citizen promulgated on 26 August.

Paramount were the political and constitutional principles; largely to be found in the Declaration. As the National

Assembly implied when it added the description 'constituent' to its title, the basic mission of the deputies was to endow France with a constitution. The Declaration was intended as a preamble to this constitution, on the model of the declarations of rights which had prefaced the constitutions of new American states like Virginia or Massachusetts a few years before. Most of its provisions were obvious products of the struggle against 'despotism' which had dominated national politics between February 1787 and August 1788. 'Ignorance, disregard, or contempt for the rights of man', says the preamble, 'are the sole cause of public misfortunes and governmental corruption,' and the whole document is an indictment of the arbitrary power wielded by the old monarchy, so inimical to the 'natural and imprescriptible rights' of liberty, property, security, and resistance to oppression (clause II). The most effective means of protecting these rights were representative institutions embodying national sovereignty and the rule of law. Few of these ideas were new. They can be traced to very diverse sources, many of them worlds away from the Enlightenment from which at first glance they may appear to stem.[2] The crucial point about them, however, is that by the late 1780s they were political commonplaces acknowledged by all literate groups. They express the broad political consensus that can be found in the *cahiers* of all three orders. The original drafting committee, indeed, tried to base its proposals quite consciously on the *cahiers*. The only item in the Declaration that gave rise to any serious disagreement was that concerning freedom of religion, which the clergy, again in full accordance with their own *cahiers* at least, were reluctant to concede without severe limitations; and they did succeed in making the relevant clauses (X, XI) sound rather grudging. Otherwise, the Declaration embodied the authentic voice of those who had made the Revolution—men of property not primarily concerned about social questions, but anxious to ensure that despotic, irresponsible government should never again bring the country to the brink of chaos.

If the Declaration was a statement of what had produced the Revolution, the decree of 11 August was largely concerned with what the Revolution had produced. Its purpose was to codify the proceedings of the night of 4 August into a formal decree. In this sense it was the National Assembly's acknowledgement of the need to conciliate the peasantry. Originating as a holding operation to prevent the total collapse of law and order, it nevertheless became a cornerstone of what are remembered as the principles of 1789, because it transformed the character of property in France. 'The National Assembly', it grandiloquently began, 'entirely destroys the feudal regime'—neglecting to note that peasant direct action had already largely accomplished this in practice. Then came the small print. The only basic feudal right to be abolished without compensation was seignorial justice. The others were more in the nature of irritants, such as various hunting rights, or institutions that were rapidly dying out, such as serfdom. Everything else was declared redeemable, and was to continue to operate until redemption. Clause VI even envisaged the possibility that new rights might subsequently come into being so long as they, too, were redeemable. Dues were, after all, property, and could not simply be destroyed. It took a special 'feudal committee' the best part of another year to work out the details of compensation procedures, and by the time they had finished—on paper at least—the peasantry were left with very little worth having from the whole business.[3] It was fortunate for them that the coercive power of successive revolutionary regimes was inadequate for making them continue to pay; and it was largely thanks to their continued resistance that the Revolution marked the end of feudalism in practice as well as in theory.

The night of 4 August was a messy occasion. In the heat of the moment people rushed to renounce or call for the renunciation of a whole range of unconnected items. Many had nothing to do with feudalism, but they were dispatched in the same session and their abolition was embodied in the same decree. Ecclesiastical tithes and other casual revenues, for instance, were abolished without compensation. This was another concession to peasant direct action, since the clergy were overwhelmingly against abolition, and only a few urban *cahiers* had called for it.[4] True, the state now assumed the burden of paying the clergy, but many deputies were not at all happy with this, and it stored up enormous problems for the future. Regional, sectional, and municipal privileges were another casualty, owing nothing this time to any need to appease the peasants. They owed little enough, either, to the demands of the *cahiers*, where the question aroused some, but far from overwhelming, interest.[5] The abolition of this jurisdictional jungle was, of course, one of the essential prerequisites for the rationalization of French government that was to be one of the Revolution's greatest achievements. But it seems to have come about as a logical corollary of the abolition of noble and clerical privileges rather than because the deputies were aware of any overwhelming need. The destruction of provincial privileges, indeed, was the very opposite of what most provinces had appeared to want, at least until the end of 1788. Until then, the best bulwark against despotism had been considered to be, if anything, a positive increase in the individual rights and liberties of the provinces.

Finally, the decree of 11 August embodied fundamental social reforms. Venality of offices was abolished, and here at least the decree was in accordance with the wishes of a large number of *cahiers*. To that majority of third estate *cahiers* which condemned venality could be added a third of noble ones.[6] No doubt the motivations of the two orders differed. The third saw venality as an obstacle to the promotion of impecunious men of talent, while nobles saw it as a means whereby the purity of their ranks was diluted by rich parvenus. The abolition was moved on 4 August by the holder of an ennobling office in the Paris parlement, and few of those in the Assembly who actually held venal offices appear to have objected to their abolition. Presumably this was because they assumed (wrongly as it proved) that they would receive full compensation.[7] But whatever the differences of motive, the result was the same. A fundamental social institution of the old order, in fact its main channel of social mobility, had been swept away. This more than anything else ensured that French society would never be the same again. Office had been the key to attaining social consequence. Henceforth social consequence was to be the key to attaining office.

But what was to be the criterion of social consequence in

the new society? The most obvious was property, a word which recurs obsessively throughout both the decree of 11 August and the Declaration. There was nothing very startling about the idea that society should be dominated by property owners. It always had been. The principle that was revolutionary, in France at least, was that the community of propertied men should not only dominate society, but exercise everyday control of government, too, through representative institutions. The commitment to representative institutions, a political rather than a social principle, was the product of a century of rising standards of education and public awareness at a time of increasing taxation and waning government prestige. Its triumph could have been foreseen from at least the 1760s. The idea that these institutions should represent a political nation of the propertied had emerged much more recently and much more suddenly. Under the old order the political nation was practically confined to the nobility, men of property certainly, but *privileged* ones. Doubts about the desirability of this could be heard from the 1770s onwards, notably in discussions over the composition of provincial assemblies. But not until 1788 was it seriously challenged, and even then not by non-privileged property owners themselves. It was the crown's decision to hold elections that first admitted the bourgeoisie to the political nation; and in this sense the fundamental social decision of eighteenth-century French history was taken not by the Revolution, but by the old order. By calling the bourgeoisie into political activity the monarchy, in its last positive act, created the élite of social 'notables', a blend of noble and non-noble property owners, that was to govern the country far into the next century. The struggles of the summer of 1789 were concerned with elaborating the precise nature of the new élite.

It was essential to establish, for instance, whether all its members had equal rights. 'Men are born and remain free and equal in rights,' says the first clause of the Declaration; 'Social distinctions may be based only on common utility.' But this did not mean that all men were equal in other ways, and the drafters of the Declaration entertained no such idea. The emphasis throughout the August decrees on the inviolability of property limited any theoretical equality almost as much as privilege had before 1789. The nobles who were now denied a privileged position in society on account of their birth, could take consolation in the fact that their property would still give them an influential part to play in the new governing class. And how, without venality, was the élite to recruit new members? It was to co-opt the talented. 'All citizens, without distinction of birth, shall be admissible to all employments', declared the decree of 11 August (clause XI), and the Declaration elaborated that they should be so admissible 'according to their capacity, and with no other distinction than that of their virtues and talents' (clause VI). These clauses were the true death-knell of noble power in France, for by them nobles lost all the social and institutional advantages upon which that power had been built. They marked the end of the privilege of birth. Privilege did not disappear; it was merely relabelled 'social distinction' and made the reward of ability and property.

Many nobles and clergy were not, of course, happy about this. It would have been extraordinary if they had been. The reluctance of most noble and clerical deputies to agree to the merger of the three orders into the National Assembly, a resistance clearly authorized by their *cahiers*, shows that the majority within the first two orders wished to retain their separate identities until a late stage, and had to be coerced into giving them up. Even so it is easy to overstate their intransigence. Only 40 per cent of noble *cahiers* had been against the merger of the orders in any circumstances. The most tangible recognition of civil equality, an equal tax burden, had already been overwhelmingly approved in their *cahiers*, so its confirmation in the Declaration (XIII) and the 11 August decree (IX) was a mere formality that nobody resisted. There was no disagreement between noble and bourgeois deputies about feudalism, which both would have preferred to retain had the peasants not forced their hands. Even careers open to the talents did not horrify nobles as much as might be expected. A quarter of their *cahiers* had called for the ennoblement of men of talent and virtue, and almost half had declared that service should be ennobled;[8] this was the principle of nobility open to talent at any rate, from which under the old order career opportunities would undoubtedly have followed. Finally, of course, there is the fact that most nobles, inside the Assembly or out, accepted all these changes. They may not have welcomed them, but most recognized that there was no future in resisting them.

Yet the nobility was deeply divided, in 1789 as beforehand. There remained an unreconciled, irreconcilable minority who preferred to emigrate or organize resistance to what had happened. It was they who got the rest a bad name, who made 'aristocracy' the word which summarized all that the Revolution was to be against, and who brought down calamity on their fellow nobles in the more dangerous years that lay ahead.

The principles of 1789, therefore, cannot be identified with the aspirations of any one of the pre-revolutionary social groups. It was not even clear, when the Estates-General met, what the principles of 1789 would turn out to be, and certainly much that had been achieved by August, and even more that was to be done later, had no mandate in the *cahiers* of that spring. Indeed, as the conservative third estate deputy, Malouet, complained, without the abrogation of mandates no revolution of any consequence could have occurred.[9] The ragged, inconsequential, coincidental, and sometimes almost haphazard way in which the principles of 1789 were formulated, is a typical enough reflection of how the Revolution itself originated.

It was neither inevitable nor predictable. What *was* inevitable was the breakdown of the old order. The ever-mounting strain of military expenditure on an inefficient financial system, together with the unwillingness of those in charge of the state to undertake any serious or at least sustained effort at structural reforms, made some sort of breakdown practically unavoidable. When it came, Louis XVI appealed to his subjects for help, but only on his own terms. By the 1780s, however, their faith in the system of government he represented had been so undermined that they refused to come to his aid except on terms of their own: the convocation of the Estates-General. It took eighteen months of bitter conflict to make the government accept this solution, and even when it did, it collapsed under pressure of renewed financial difficulties rather than the strength of the opposition.

The government's collapse left a vacuum of power. The forces of opposition were taken by surprise by it. They had known well enough what they were against, which could be summarized in the word 'despotism'. They knew, too, that the remedy was the Estates-General. Beyond that, however, nobody was clear what he wanted, and the winter of 1788/9 was spent in impassioned discussion of this question. By the time the Estates actually met, some degree of political consensus had emerged. All parties agreed on the need for a constitution embodying a representative form of government with guarantees for individual liberty and the rule of law. All agreed, too, that fiscal privileges should have no place in the new order and that the time was ripe for a complete overhaul of the systems of finances, administration, and justice. All this the Revolution was to bring about; but not until other problems, unforeseen eighteen months before, had first been settled.

First among these problems came the integration of the bourgeoisie into the political nation. Until 1788, it had merely been an increasingly well-informed spectator of public affairs, with no evident desire to participate. But the calling of elections, and indeed the very composition of the Estates-General, inevitably drew it into activity, and it found itself wooed for its potential political weight. By the spring of 1789 it had conceived the desire, in Sieyès's phrase, 'to be something'. The nobility was called upon to share political power in the new order. Unfortunately, the claim to share political power, which few nobles would have dismissed out of hand, became confused with attacks on their age-old social leadership, and by the time of the elections there was deep mutual suspicion and misunderstanding. The *cahiers*, by forcing people to transform vague feelings into words, compounded growing antagonisms. By the time the Estates actually met, nobility and bourgeoisie, who basically agreed on so much, had become competitors for power rather than partners in its exercise. Only when the right of nobles to separate treatment in anything had been destroyed could the new political élite of propertied 'notables', an amalgam of former nobles and bourgeois, get down to the exercise of power.

This took until mid-July, and it might never have come about but for the coincidence of the economic crisis, which brought the populace into political activity. In the midst of the worst economic difficulties for generations, the ordinary people of France were led to expect unforeseen relief from the Estates-General. They, too, had their vision clarified by the *cahiers*. But when relief did not come they grew impatient and began to take the law into their own hands. They also accepted the claims of third estate deputies, their own deputies after all, that aristocratic obstructionism was responsible for the delay. When the government, bestirring itself at last, appeared poised to throw its weight behind those who were impeding progress, the people of Paris intervened to save the new-born National Assembly from destruction. Few deputies felt happy to have such an ungovernable ally, but when troops closed in on Paris and Versailles there seemed to be no alternative. Even then it is not clear that Paris could have saved the situation if the king had not lost faith in his own army. But the matter was not put to the test; the Bastille fell; and Louis XVI capitulated. Power in France now lay unchallengeably with a propertied élite recognizing no special place for nobles, either at national or local level.

The National Assembly at last set about constitution making; but the ramifications of the economic crisis were not yet at an end. The peasantry as well as the people of Paris had been thrown into turmoil by the coincidence of economic adversity and political excitement, and they took advantage of uncertain times to throw off the most dispensable of their burdens, the 'feudal' dues. The deputies were appalled at such wanton disregard for property, but there was no prospect of restoring order in the countryside without accepting what the peasants had done. And so, in spite of themselves, the men of 1789 acquiesced in the destruction of feudalism.

Only now could France's new ruling élite begin to assess what they stood for, and what they had achieved. As victors will, they soon convinced themselves that all had gone according to plan from the start. But there was no plan, and nobody capable of making one, in 1787. Nobody then could have predicted that things would work out as they did. Hardly anybody would have felt reassured if they could. For the French Revolution had not been made by revolutionaries. It would be truer to say that the revolutionaries had been created by the Revolution.

Notes

Abbreviations used in the notes

A.H.R.	*American Historical Review*
A.H.R.F.	*Annales historiques de la Révolution française*
Annales	*Annales: Économies, Sociétés, Civilisations*
Econ.H.R.	*Economic History Review*
E.H.R.	*English Historical Review*
Fr.Hist.St.	*French Historical Studies*
J.Econ.Hist.	*Journal of Economic History*
J.M.H.	*Journal of Modern History*
P. and P.	*Past and Present*
R. d'Hist. Mod. et Contemp.	*Revue d'Histoire Moderne et Contemporaine*

Note to Introduction

1 L. Stone, *The Causes of the English Revolution, 1529–1642* (London, 1972), 26.

Notes to Part I

1 G. Lefebvre, *The Coming of the French Revolution* (paperback edn., New York, 1957), Introduction.

2 For a survey of recent Marxist writings on this subject, see G. Ellis, 'The "Marxist Interpretation" of the French Revolution', *E.H.R.* xc iii (1978), 353–76.

3 *La Révolution française* (*Peuples et Civilisations*, vol. XIII) (Paris, 1951), 1–2, 113–53. Eng. trans. (London and New York, 1962–4).

4 Notably, *Les Paysans du Nord pendant la Révolution française* (2 vols., Lille, 1924), (new edn., Paris 1972), and *La Grande Peur de 1789* (Paris, 1932), Eng. trans. (London, 1973). On the personality and influence of Lefebvre, see the commemorative issue of *Annales historiques de la Révolution française* xxxii (1960), and the sketch in R.C. Cobb, *A Second Identity. Essays on France and French History* (Oxford, 1969).

5 C.E. Labrousse, *Esquisse du mouvement des prix et des revenus en France au XVIIIe siècle* (2 vols., Paris, 1933); *La Crise de l'économie française à la fin de l'Ancien Régime et au début de la Révolution* (Paris, 1944).

6 P. Gaxotte, *La Révolution française* (Paris, 1928), Eng. trans. (London, 1932); B. Faÿ, *La Grande Révolution* (Paris, 1959); F. Braesch, *1789: L'Année cruciale* (Paris, 1941).

[7] For the early impact of Lefebvre, see B.F. Hyslop's review of Palmer's translation in *A.H.R.* liii, (1947/8), 808–10, and her earlier article, 'Recent Work on the French Revolution', ibid. (1942), 489–90.

[8] e.g. A. Goodwin, 'Calonne, the Assembly of French Notables of 1787 and the Origins of the *Révolte Nobiliaire*', *E.H.R.* lxi (1946).

[9] A. Goodwin, *The French Revolution* (London, 1953), chs. I–V.

[10] A. Soboul, *Précis d'histoire de la Révolution française* (Paris, 1962), Eng. trans. (2 vols., London, 1975).

[11] F. Furet and D. Richet, *La Révolution française* (2 vols., Paris, 1965), Eng. trans. (London, 1970).

[12] A. Cobban, *The Myth of the French Revolution* (London, 1955), conveniently reprinted in his *Aspects of the French Revolution* (London, 1968), 90–111.

[13] *Aspects*, 106.

[14] *French Revolution*, 53–72.

[15] *A History of Modern France* (London, 1957), I, 140, 153, 167, 257.

[16] Review of *The Myth of the French Revolution* in *A.H.R.F.* xxviii, (1956), 337–45.

[17] At least, according to P. Dawson, *Provincial Magistrates and Revolutionary Politics in France, 1789–1795* (Cambridge, Mass., 1972), 11.

[18] M. Reinhard, 'Sur l'histoire de la Révolution française. Travaux récents et perspectives'. *Annales*, xiv, (1959), 557–62.

[19] e.g. M. Vovelle and D. Roche, 'Bourgeois, rentiers, propriétaires: éléments pour la définition d'une catégorie sociale à la fin du XVIIIe siècle', *Actes du 84e Congrès national des sociétés savantes (Dijon, 1959), Section d'histoire moderne et contemporaine* (Paris, 1962), 483–512, Eng. trans. in J. Kaplow (ed.), *New Perspectives on the French Revolution. Readings in Historical Sociology* (New York, 1965), 25–46; A. Daumard and F. Furet, *Structures et relations sociales à Paris au milieu du XVIIIe siècle* (Paris, 1961), 35–7, 39–40, 92.

[20] Elinor G. Barber, *The Bourgeoisie in 18th Century France* (Princeton, N.J., 1955).

[21] (Cambridge, 1964).

[22] *Social Interpretation*, 67.

[23] Ibid., 172–3.

[24] Ibid., 23.

[25] *Études orléanaises, i. Contribution à l'étude des structures sociales à la fin du XVIIIe siècle* (Paris, 1962).

[26] Godechot's review was in *Revue historique* ccxxxv (1966), 205–9; the *A.H.R.F.* did not deign to notice the book at all. Cobban's replies are most conveniently found in his *Aspects*, 264–87. A full account of the controversy can be found, with copious references, in ch. I of Dawson, *Provincial Magistrates*. See also Gerald C. Cavanaugh, 'The Present State of French Revolutionary

Historiography: Alfred Cobban and Beyond', *French Historical Studies* (1972).

27 See Stone, *Causes of the English Revolution*, ch. II.

28 *Irish Historical Studies* (1964–5) xiv, 192.

29 *A Social History of the French Revolution* (London, 1963), 20–2, 63–4, 80–1.

30 See his essay, 'The Enlightenment and the French Revolution', reprinted in *Aspects*, 18–28.

31 See 'The *Parlements* of France in the Eighteenth Century', *Aspects*, 68–82; *History of Modern France*, I, 253–4.

32 'France' in A. Goodwin (ed.), *The European Nobility in the Eighteenth Century* (London, 1953), 28.

33 'L'Aristocratie parlementaire française à la fin de l'Ancien Régime', *Revue historique* ccviii (1952), 1–14.

34 Subtitled *The Regrouping of the French Aristocracy after Louis XIV* (Cambridge, Mass., 1953).

35 *The Nobility of Toulouse in the Eighteenth Century* (Baltimore, Md., 1960), 177.

36 'The Noble Wine Producers of the Bordelais in the Eighteenth Century', *Econ. H.R.* xiv (1961), 18–33; 'The Provincial Noble: A Reappraisal', *A.H.R.* lxviii (1962–3).

37 'Nobles, Privileges and Taxes in France at the end of the Ancien Régime', *Econ.H.R.* xv (1962/3), 451–75. Her arguments have not gone unchallenged; see G.J. Cavanaugh, 'Nobles, Privileges and Taxes in France. A Revision Reviewed', *Fr. Hist. St.* (1974), viii 681–92. Cavanaugh's criticisms, which seem valid, nevertheless do not diminish the importance of Behrens's ideas in opening up the question of nobility.

38 *The Ancien Régime* (London, 1967), 46–62.

39 'Types of Capitalism in Eighteenth Century France', *E.H.R.* lxxix (1964), 478–97.

40 'Noncapitalist Wealth and the Origins of the French Revolution', *A.H.R.* lxxii (1967), 469–96.

41 Ibid., 487.

42 For an introduction to this debate, see J.H.M. Salmon, 'Venal Office and Popular Sedition in 17th Century France', *P. and P.* 37 (1967).

43 See R. Mousnier, *Les Hiérarchies sociales de 1450 à nos jours* (Paris, 1969), Eng. trans., *Social Hierarchies* (1973).

44 R. Mousnier, *La Société française de 1770 à 1789* (Paris, 1970), 1–2; see, too, his debate with Labrousse and Soboul in *L'Histoire sociale: sources et méthodes* (Paris, 1967), 26–31.

45 e.g. S. Pillorget, *Apogée et déclin des sociétés d'ordres 1610–1787* (Paris, 1969).

46 e.g. P. Goubert, *L'Ancien Régime I: La Société* (Paris, 1969), 235. Eng. trans. (1973).

47 In his survey of the historiography of the Revolution, *Un jury pour la Révolution* (Paris, 1974), Godechot nowhere mentions Cobban or Taylor.

[48] C. Mazauric, *Sur la Révolution française* (Paris, 1970).

[49] F. Furet, 'Le catéchisme révolutionnaire', *Annales*, xxvii (1971), 255–89, reprinted in his *Penser la Révolution française* (Paris, 1978), trans. Elborg Forster as *Interpreting the French Revolution* (Cambridge, 1981).

[50] D. Richet, 'Autour des origines idéologiques lointaines de la Révolution française: Élites et despotisme', *Annales*, xxiv (1969), 1–23. His ideas were further elaborated in *La France moderne: l'esprit des institutions* (Paris, 1973).

[51] 'Noncapitalist Wealth', 491.

[52] 'Revolutionary and Nonrevolutionary Content in the *Cahiers* of 1789: An Interim Report', *Fr. Hist. St.*, vii (1972), 479–502.

[53] G. Richard, *Noblesse d'affaires au XVIIIe siècle* (Paris, 1974).

[54] G. Chaussinand-Nogaret, *La Noblesse au XVIIIe siècle. De la féodalité aux lumières* (Paris, 1976), 119–61.

[55] O.H. Hufton, *Bayeux in the Later Eighteenth Century. A Social Study* (Oxford, 1967), 42, 57–80; J. Sentou, *Fortunes et groupes sociaux à Toulouse sous la Révolution* (Toulouse, 1969), 146; M. Garden, *Lyon et les Lyonnais au XVIIIe siècle* (Paris, 1970), 358, 360, 362, 387–98.

[56] J.P. Poussou in G. Pariset (ed.), *Bordeaux au XVIIIe siècle* (Bordeaux, 1968), 351–2.

[57] Y. Durand, *Les Fermiers-Généraux au XVIIIe siècle* (Paris, 1971), 294–301.

[58] Chaussinand-Nogaret, *La Noblesse au XVIIIe siècle*, 48.

[59] Everything depends on how many nobles there were. Chaussinand-Nogaret, ibid., 48, argues that there were only 110,000 to 120,000; J. Meyer, 'La Noblesse française au XVIIIe siècle: aperçu des problèmes', *Acta Poloniae Historica* xxxvi (1977), 18–22, prefers the much higher figure of 350,000. Both agree, however, that the number of new nobles was far higher than previously thought.

[60] For a survey of this evidence, see W. Doyle, 'Was There an Aristocratic Reaction in Pre-Revolutionary France?' *P. and P.* 57 (1972), reprinted in D. Johnson (ed.), *French Society and the Revolution* (Cambridge, 1976), 3–20; of fundamental importance, too, is D.D. Bien, 'La Réaction aristocratique avant 1789: l'exemple de l'armée' *Annales*, xxix, (1974).

[61] L.R. Berlanstein, *The Barristers of Toulouse in the Eighteenth Century (1740–1793)* (Baltimore, Md., 1975), 34–5, 51–5.

[62] Ibid; 149–60, 177–82; Dawson, *Provincial Magistrates*, 152–84, 200–40.

[63] J. Meyer, *La Noblesse Bretonne au XVIIIe siècle* (2 vols., Paris, 1966).

[64] Bien, 'La Réaction'.

[65] Bailey Stone, 'Robe against Sword: The Parlement of Paris and the French Aristocracy, 1774–1789', *Fr.Hist.St.* ix (1975).

[66] Chaussinand-Nogaret, *La Noblesse au XVIIIe siècle*, 181–226.

[67] 'Nobles, Bourgeois and the Origins of the French Revolution',

P. and P. 60 (1973), reprinted in Johnson (ed.), *French Society and the Revolution*, 88–131. All references are to this edition.

68 Johnson, 95.

69 Lefebvre, *The Coming*, 30–4.

70 See E.L. Eisenstein, 'Who Intervened in 1788? A Commentary on *The Coming of the French Revolution*', *A.H.R.* lxxi, (1965), 77–103, and the rejoinders of J. Kaplow and G. Schapiro, *A.H.R.* lxxii (1967), 498–522.

71 Lucas in Johnson, *French Society*, 128.

72 A typical reaction of reserve is that of Betty Behrens in a long review article, 'The Ancien Régime and the Revolution', *Historical Journal* xvii (1974), 630–43.

73 W. Doyle, 'The Price of Offices in Pre-Revolutionary France', *Historical Journal*, 27 (1985).

74 *Une histoire des Élites, 1700–1848* (Paris/The Hague, 1975).

75 For careful analyses of the mental processes by which these views emerged, see J.M. Roberts, 'The Origins of a Mythology: Freemasons, Protestants and the French Revolution', *Bull.I.Hist.R.* xliv (1971), 78–97; *The Mythology of the Secret Societies* (1972); and 'The French Origins of the "Right"', *Transactions of the Royal Historical Society* xxiii (1973), 27–53.

76 e.g. B. Faÿ, *La Franc-Maçonnerie et la révolution intellectuelle du XVIIIe siècle* (Paris, 1935).

77 *Les Origines intellectuelles de la Révolution française 1715–1787* (Paris, 1933).

78 For Mornet's methods, see his 'Les Enseignements des bibliothèques privées, 1750–1780', *Revue d'Histoire Littéraire de la France* July–Sept. (1910). For his legacy, see G. Bollême, A. Dupront, *et al*, *Livre et société dans la France du XVIIIe siècle* (Paris/The Hague, 1965).

79 *The Coming*, 4, 42–4.

80 *Aspects*, 18–28.

81 *History of Modern France*, 96–109.

82 J. McDonald, *Rousseau and the French Revolution 1762–1791* (London, 1965), 50.

83 Thus Goodwin, *The French Revolution*, hardly mentions the Enlightenment. M.J. Sydenham, *The French Revolution* (London, 1965), 20, 23–4, more or less follows Lefebvre and Mornet, as if disinclined to explore the minefield independently.

84 N. Hampson, *The First European Revolution 1776-1815* (1969), 9–40; 'The "Recueil des pièces intéressantes pour servir à l'histoire de la Révolution en France" and the Origins of the French Revolution', *Bulletin of the John Rylands Library* xlvi (1964); and *The French Revolution: a Concise History* (1975), 37–54.

85 See the devastating review of *Rousseau and the French Revolution* by R.A. Leigh in *Historical Journal* xii (1969), 561–3.

86 Dawson, *Provincial Magistrates*, 115–25; Berlanstein, *Barristers of Toulouse*, 93–122.

[87] *Les Origines*, 453-65.

[88] *L'Ancien Régime* (trans., Patterson, Oxford, 1956) 152-3.

[89] 'Revolutionary and Nonrevolutionary Content', 494.

[90] *La Noblesse au XVIIIe siècle*, chs. I, VII, and VIII.

[91] D. Richet, 'Autour des origines'.

[92] R. Darnton, *The Business of Enlightenment. A Publishing History of the Encyclopédie, 1775-1800* (Cambridge, Mass., 1979).

[93] L.W.B. Brockliss, *French Higher Education in the Seventeenth and Eighteenth centuries* (Oxford, 1987).

[94] K.M. Baker, 'French Political Thought at the Accession of Louis XVI', *J.M.H.* Vol. L (1978), 278-303; and 'On the Problem of the Ideological Origins of the French Revolution', in D. La Capra and S.L. Kaplan (eds.), *Modern European Intellectual History. Reappraisals and New Perspectives* (Ithaca and London, 1983), 197-219.

[95] e.g. A. Le Bihan, *Loges et chapitres de la Grande Loge et du Grand Orient de France (seconde moitié XVIIIe siècle)* (Paris, 1968); M. Agulhon, *Pénitents et Francs-Maçons dans l'ancienne Provence* (Paris, 1968); P. Chevallier, *Histoire de la Franc-Maçonnerie française, I; la Maçonnerie: école de l'égalité 1725-1799* (Paris, 1974). R. Halévi, *Les loges maconniques dans la France d'Ancien Regime, Aux origines de la sociabilité démocratique* (Paris, 1984).

[96] R. Darnton, *Mesmerism and the End of the Enlightenment in France* (Cambridge, Mass., 1968).

[97] 'The High Enlightenment and the Low Life of Literature in Pre-Revolutionary France', *P. and P.* 51 (1971), reprinted in Johnson, *French Society*, 53-87.

[98] See above, n. 5.

[99] 'Le Mouvement des prix et les origines de la Révolution française', *Annales d'histoire économique et sociale* (1937), reprinted in *Études sur la Révolution française* (Paris, 1963), 197-237, esp. p. 223. However, the American historian D.S. Landes, in 'The Statistical Study of French Crises', *J. Econ. History* x (1950), 195-211, argues that some of Labrousse's figures are made to bear a heavier burden of interpretation than they can carry. He also suggests that the economic theories underlying his approach ignore certain ambiguities indicating that the crisis may have been less acute in industry than Labrousse believed.

[100] '1848-1830-1789: How Revolutions Are Born', a paper first delivered to the *Congrès historique du centenaire de la Révolution de 1848* in 1948, and reprinted and translated in F. Crouzet, W.H. Chaloner, and W.M. Stern, *Essays in European Economic History 1789-1914* (1969), p. 9.

[101] J.F. Bosher, *French Finances 1770-1795: From Business to Bureaucracy* (Cambridge, 1970), ch. X, esp. 190-1; 'The French Crisis of 1770', *History* lvii (1972); 17-30.

[102] J.C. Toutain, 'Le Produit de l'agriculture française de 1700 à

1958', *Cahiers de l'Institut de Science Économique appliquée* (1961), esp. 276–7.

103 M. Morineau, *Les Faux-Semblants d'un démarrage économique. Agriculture et démographie en France au XVIIIe siècle* (Paris, 1970. *Cahiers des Annales* 30).

104 So far nobody has disputed Morineau's conclusions on the same scale as he presented them. But Emmanuel Le Roy Ladurie, 'L'Histoire immobile', *Annales* xxix (1974), 691, declares himself unconvinced by Morineau's demographic arguments, and therefore seems to imply (without analysing the matter) that there *must* have been some dramatic rise in agricultural productivity after all to permit the population rise.

105 J. Dupâquier, *La Population française aux xviie et xviiie siècles* (Paris, 1979).

106 *The Economic Modernisation of France (1730–1880)* (London, 1975).

107 Doyle, 'Was There an Aristocratic Reaction?'

108 Simiand was Labrousse's master. His classification of price history into 'A' phases (upward movements) and 'B' phases (deceleration or reversal) has become classic, as has his general delineation of these phases between the sixteenth and nineteenth centuries, outlined in his *Recherches anciennes et nouvelles sur le mouvement général des prix du XVIe au XIXe siècle* (Paris, 1932).

109 Note, in this context, the telling title of one of the earliest, and best, nineteenth-century surveys of the politics of Louis XVI's reign: F.X.J. Droz, *Histoire du règne de Louis XVI pendant les années où l'on pouvoit prévenir ou diriger la Révolution française* (3 vols., Paris, 1839–42).

110 See the article 'Parlements' in his still-influential *Dictionnaire des institutions de la France aux XVIIe et XVIIIe siècles* (Paris, 1923), and his biographical studies of major opponents of the parlements, e.g. *Machault d'Arnouville* (Paris, 1891), *La Bretagne et le duc d'Aiguillon* (Paris, 1898), and *Le Garde des Sceaux Lamoignon et la réforme judiciaire de 1788* (Paris, 1905).

111 See above, n. 6; and F. Piétri, *La Réforme de l'État au XVIIIe siècle* (Paris, 1935).

112 *The Coming*, 14–21.

113 H. Méthivier, *L'Ancien Régime* (Paris, 1961), and *La Fin de l'Ancien Régime* (Paris, 1970).

114 L. Laugier, *Un ministère réformateur sous Louis XV: Le Triumvirat* (Paris, 1975), with an introduction by Pierre Gaxotte.

115 M. Antoine, *Le Conseil du Roi sous le règne de Louis XV* (Paris, 1970). See especially the conclusion, 633–4.

116 *A History of Modern France*, I, 91–7, and 'The *Parlements* of France in the Eighteenth Century', *Aspects*, 68–82.

117 E. Faure, *12 Mai 1776. La Disgrâce de Turgot* (Paris, 1961).

118 *A History*, 95–6.

119 J. Egret, *Le Parlement de Dauphiné et les affaires publiques dans*

la deuxième moitié du XVIIIe siècle (2 vols., Paris and Grenoble, 1942).

120 J. Egret, *Louis XV et l'opposition parlementaire 1715-1774* (Paris, 1970).

121 J.H. Shennan, *The Parlement of Paris* (London, 1968).

122 W. Doyle, 'The Parlements of France and the Breakdown of the Old Regime, 1770-1788', *Fr. Hist. St.* vi (1970).

123 See D.C. Hudson, 'In Defense of Reform: French Government Propaganda during the Maupeou Crisis', *Fr. Hist. St.* viii (1973), 53, or P. Alatri, *Parlamenti e lotta politica nella Francia del '700* (Bari, 1977), 401.

124 *The Coming*, 19-25.

125 W.J. Pugh, 'Calonne's "New Deal"', *J.M.H.* ix (1939), 289 312.

126 A. Goodwin, 'Calonne', 202-34 and 329-77.

127 Ibid., 364.

128 J. Egret, *La Pré-Révolution française 1787-1788* (Paris, 1962), trans. by W.D. Camp (Chicago, 1977). See especially, in the French edn., chs. I, III, and pp. 306-15, 369-72; see also Bosher, *French Finances*, ch. II.

129 See Lefebvre's rather grudging portrait, *The Coming*, 49-50.

130 E. Lavaquery, *Necker, fourrier de la Révolution* (Paris, 1933), holds him blinded by vanity; E. Chapuisat, *Necker (1732-1804)* (Paris, 1938), sees him as a man of goodwill, but overwhelmed by the circumstances in which he found himself.

131 Bosher, *French Finances*, ch. VIII; J. Egret, *Necker, ministre de Louis XVI* (Paris, 1975), 104-26.

132 H. Grange, *Les Idées de Necker* (Paris, 1974): rather an uncritical study, but a useful and interesting survey for all that.

133 R.D. Harris, 'Necker's *Compte rendu* of 1781: A Reconsideration' *J.M.H.* xlii (1970), and 'French Finances and the American War, 1777-83', *J.M.H.* xlviii (1976). Harris has now produced two full-scale studies of Necker's ministries: *Necker, Reform Statesman of the Ancien Regime* (Berkeley, Calif., 1979), *Necker and the Revolution of 1789* (Lanham, 1986).

134 See R.R. Palmer, *The Age of the Democratic Revolution. I. The Challenge* (Princeton, N.J., 1959), or J. Godechot, *Les Révolutions 1770-1799* (Paris, 1963), trans. as *France and the Atlantic Revolution of the 18th Century* (London, 1965).

135 P. Girault de Coursac, *L'Éducation d'un roi. Louis XVI* (Paris, 1972).

Notes to Chapter 1

1 Harris, 'Necker's *Compte rendu*' and 'French Finances and the American War' show admirably the difficulty and complexity of arriving at reliable conclusions; see, too, Bosher, *French Finances*, chs. II-IV.

2 See Egret, *La Pré-Révolution*, 5-6; Goodwin, 'Calonne', 212, 336.

[3] That is to say, the right to collect and dispose of that amount of revenue in 1787 had been sold off for ready cash.
[4] Bosher, *French Finances*, 4.
[5] J.C. Riley, *The Seven Years War and the Old Regime in France, The Economic and Financial Toll* (Princeton, 1986), 191.
[6] J. Necker, *A Treatise on the Administration of the Finances of France* (London, 1785) II, 505—9. Out of a total expenditure of 610 millions, these items made up 87 millions. See too Harris, 'French Finances', 245-6.
[7] D. Dakin, *Turgot and the Ancien Régime in France* (London, 1939), 156.
[8] See the table in Palmer, *Age of the Democratic Revolution*, I, 155.
[9] F. Hincker, *Les Français devant l'impôt sous l'Ancien Régime* (Paris, 1971), 40-1. See, too, J.C. Riley, 'French Finances, 1727-1768', *J.M.H.* lix (1987).
[10] Cavanaugh, 'Nobles, Privileges and Taxes in France. A Revision Reviewed'.
[11] See Riley, *Seven Years War*, ch. 6.
[12] J.F. Bosher, 'The French Crisis of 1770'.
[13] See below, pp. 74-5.
[14] G.V. Taylor, 'The Paris Bourse on the Eve of the Revolution, 1781-1789', *A.H.R.* lxvii (1962), 956-7.
[15] Dakin, *Turgot*, 175.
[16] Harris, 'French Finances', 240.
[17] J.C. Riley, 'Dutch investment in France, 1781-1787', *J.Econ. Hist.* xxxiii (1973), 732-43.
[18] Harris, 'Necker's *Compte rendu*', *passim*; Egret, *Necker*, 169-75.
[19] Quoted in Goodwin, 'Calonne', 206.
[20] Taylor, 'Paris Bourse', 963, n. 38; Harris 'French Finances', has revised the figure for Necker's loans. It may be that that for Calonne's should be re-examined, too.
[21] Bosher, *French Finances*, 190-1.
[22] This paragraph is based entirely on ibid., 3-196, *passim*.
[23] Quoted in Egret, *Pré-Révolution*, 6.
[24] Ibid., 6-7; Goodwin, 'Calonne', 209-10.

Notes to Chapter 2

[1] P. Grosclaude, *Malesherbes, témoin et interprètre de son temps* (Paris, 1961), 279.
[2] M. Antoine, *Le Conseil du roi*, 599-627, paints a sympathetic and largely convincing picture of Louis XV in the performance of his kingly duties.
[3] See Girault de Coursac, *L'Éducation d'un roi* for Louis's youth and character.
[4] Antoine, *Le Conseil du roi*, 610.
[5] Grange, *Les Idées de Necker*, 34-52, 359-99; Harris, *Necker*, chs. vi-xiv.

[6] Egret, *Necker*, 171-9.

[7] Doyle, 'The Parlements of France', 454-6.

[8] Egret, *Necker*, 186-212; Lavaquery, *Necker*, 221-74.

[9] Antoine, *Le Conseil du roi*, is now the authoritative work on this subject, superseding all previous accounts.

[10] Under Louis XIV this body had been known as the *conseil d'en haut*, but the usage lapsed in the eighteenth century. See Antoine, *Le Conseil du roi*, 122.

[11] Necker, a Protestant, was technically debarred by his religion from the post of comptroller-general, which was occupied by a straw man; but it was he who really managed the finances between 1777 and 1781.

[12] There were in fact two other councils, finances and commerce, whose spheres of competence speak for themselves. But they were increasingly adjuncts of the comptroller-general's department, not the source of major policy decisions, and met less and less often. See Antoine, *Le Conseil du roi*, 132-9.

[13] Antoine, *Le Conseil du roi*, 223-37; V.R. Gruder, *The royal provincial Intendants: A Governing Élite in eighteenth century France* (Ithaca, N.Y., 1968), 71-94.

[14] The number of generalities varied from time to time, but there were thirty-four in 1789. See M. Bordes, *L'Administration provinciale et municipale en France au XVIIIe siècle* (Paris, 1972), 130-2; pp. 116-59 of this work are the best brief summary of the functioning of the intendants.

[15] M. Bordes, 'Les Intendants de Louis XV', *Revue historique* (1960), and 'Les Intendants éclairés de la fin de l'Ancien Régime', *Revue d'histoire économique et sociale* (1961).

[16] Rathéry (ed.), *Journal et mémoires du Marquis d'Argenson* (Paris, 1859), I, 43.

[17] M. Bordes, 'Les Intendants de province aux XVIIe et XVIIIe siècles', *L'Information historique* 30 (1968) 112-6.

[18] H. Fréville, *L'Intendance de Bretagne 1689-1790* (3 vols., Rennes, 1953) III, 132-7.

[19] W. Doyle, *The Parlement of Bordeaux and the End of the Old Regime 1771-1790* (London, 1974), 60.

[20] See Bordes, *L'Administration provinciale*, 67-115, for a full account of the estates and their struggle against the *élus*.

[21] Walloon Flanders, Artois, and Cambrésis, in the north; Quatre Vallées, Pays de Marsan, Pays de Labourd, Pays de Soule, Nébouzan, Foix, Basse-Navarre, Bigorre, and Béarn in the Pyrenees; and (from its annexation in 1768), Corsica.

[22] See Fréville, *L'Intendance de Bretagne*, II, 69-342; A. Rebillon, *Les États de Bretagne de 1661 à 1789* (Paris/Rennes, 1932), 338-69.

[23] N. Temple, 'The Control and Exploitation of French Towns during the Ancien Regime', *History* li (1966), reprinted in R.F. Kierstead (ed.), *State and Society in Seventeenth Century France* (New York, 1975), 67-93.

[24] At least where these offices were not venal, or where venal ones had been bought up by municipalities under government fiscal blackmail. This process is well outlined in the articles by Temple, cited above, n. 23.

[25] M. Bordes, *La Réforme municipale du contrôleur-général Laverdy et son application (1764–1771)* (Toulouse, 1968), summarized in Bordes, *L'Administration provinciale*, 199–345.

[26] See N. Temple, 'Municipal Elections and Municipal Oligarchies in Eighteenth Century France' in J.F. Bosher (ed.), *French Government and Society 1500–1850: Essays in Memory of Alfred Cobban* (London, 1973), 70–91, for an illuminating exploration of these problems.

[27] O.H. Hufton, *The Poor of Eighteenth-Century France 1750–1789* (Oxford, 1974), 221–3; I.A. Cameron, 'The Police of Eighteenth Century France', *European Studies Review* vii (1977).

[28] Garden, *Lyon et les Lyonnais*, 39, 524.

[29] A. Williams, *The Police of Paris, 1718–1789* (Baton Rouge, La., 1979), 62–6; see also J. Godechot, *The Taking of the Bastille. July 14th 1789* (London, 1970), 78–84; for the lieutenant's duties see R. Darnton, 'The Memoirs of Lenoir, Lieutenant de Police of Paris, 1774–1785', *E.H.R.* lxxxv (1970), 537–8.

[30] See R. Pillorget, 'Les Problèmes du maintien de l'ordre public en France, entre 1774 et 1789', *L'Information historique* 39 (1977) 114–19.

[31] G. Pagès, *La Monarchie d'Ancien Régime en France* (Paris, 1928), 182–91.

Notes to Chapter 3

[1] J.M. Hayden, *France and the Estates General of 1614* (Cambridge, 1974).

[2] L. Rothkrug, *Opposition to Louis XIV. The Political and Social Origins of the French Enlightenment* (Princeton, N.J., 1965), 133–4.

[3] M. Marion, *Dictionnaire des institutions de la France aux XVIIe et XVIIIe siècles* (Paris, 1923), 216–17.

[4] J. Flammermont, *Le Chancelier Maupeou et les parlements* (Paris, 1883), 256, 261, 271–2.

[5] Goodwin, 'Calonne', 225–6.

[6] See above, pp. 61–2.

[7] Pillorget, 'Maintien de l'ordre public', 116; see also E. Le Roy Ladurie, 'Révoltes et contestations rurales en France de 1675 à 1788', *Annales* xxix (1974), for speculations on the causes of the decline in popular uprisings in the eighteenth century.

[8] G. Rudé, *The Crowd in History. A Study of Popular Disturbances in France and England 1730–1848* (New York, 1964), 22–32.

[9] S.L. Kaplan, *Bread, Politics and Political Economy in the Reign of Louis XV* (2 vols., The Hague, 1976), I, 200–14; II, 444–50.

[10] Full sessions convened in practice only every ten years.

11 The most convenient recent treatment of these matters is L.S. Greenbaum, *Talleyrand, Statesman-Priest. The Agent-General of the Clergy and the Church of France at the end of the Old Regime* (Washington, D.C., 1970), 24-8, 37-40.

12 Ibid., 56. 56.

13 Marion, *Machault d'Arnouville*, 198-260, 313-29.

14 Greenbaum, *Talleyrand*, 88-92.

15 Parlements were established at Paris, Toulouse, Grenoble, Bordeaux, Dijon, Rouen, Aix, Rennes, Pau, Metz, Besançon, Douai, and Nancy; there were also four 'sovereign councils' established at Arras, Colmar, Perpignan, and Ajaccio.

16 These included the four sovereign councils: the Grand Conseil; the Cour des Monnaies; four Cours des Aides (Paris, Bordeaux, Montauban, Clermont); and eleven Chambres des Comptes (Paris, Aix, Grenoble, Dijon, Nantes, Rouen, Montpellier, Pau, Dôle, Bar, and Nancy).

17 Ford, *Robe and Sword*, Bks. I and II, give a convenient survey of this group, along with all necessary definitions, in the earlier eighteenth century.

18 Gruder, *Royal provincial Intendants*, 34-51; Antoine, *Conseil du roi*, 223-37; F. Bluche, 'L'Origine sociale du personnel ministériel français au XVIIIe siècle', *Bulletin de la Société d'Histoire moderne* xii (1957), 9-13.

19 R. Villers, *L'Organisation du parlement de Paris et des conseils supérieurs d'après la réforme de Maupeou* (Paris, 1937). And the plan was of course abandoned when the old parlements were restored in 1774.

20 See Taylor, 'Noncapitalist Wealth', 477, and Doyle, 'Price of Offices', 848, 859-60.

21 M. Lhéritier. *L'Intendant Tourny (1695-1760)* (2 vols., Paris, 1920), II, 506-21 for the case of Tourny; Doyle, *Parlement of Bordeaux*, 231-44, for the case of Dupré de Saint-Maur.

22 A.L. Moote, *The Revolt of the Judges. The Parlement of Paris and the Fronde 1643-1652* (Princeton, N.J., 1971); A.N. Hamscher, *The Parlement of Paris after the Fronde 1653-1673* (Pittsburgh, Pa., 1976).

23 For an introduction to this question, see A.L. Moote, 'Law and Justice under Louis XIV', in J.C. Rule (ed.), *Louis XIV and the Craft of Kingship* (Columbus, Ohio, 1969), 229-35.

24 J.H. Shennan, 'The Political Role of the Parlement of Paris 1715-1723', *Historical Journal* viii (1965), and 'The Political Role of the *Parlement* of Paris under Cardinal Fleury', *E.H.R.* lxxxi (1966); J.D. Hardy, Jr., *Judicial Politics in the Old Regime. The Parlement of Paris during the Regency* (Baton Rouge, La., 1967); J. Egret, *Louis XV et l'opposition*, 9-49.

25 Doyle, 'Parlements', 454-6.

26 D. Van Kley, *The Jansenists and the Expulsion of the Jesuits from France 1757-1765* (New Haven, Conn., and London, 1975).

[27] Egret, *Louis XV et l'opposition*, 68-92.
[28] Ibid., 72-6.
[29] D.C. Hudson, 'The Parlementary Crisis of 1763 in France and Its Consequences', *Canadian Journal of History* vii (1972), 108; Egret, *Louis XV et l'opposition* 179-80.
[30] Egret, *Louis XV et l'opposition*, 137-9.
[31] Kaplan, *Bread, Politics and Political Economy*, II, 451-71.
[32] Egret, *Louis XV et l'opposition*, 170-3.
[33] Doyle, 'Parlements', 416-22.
[34] The best account is still Flammermont, *Le Chancelier Maupeou et les parlements.* Laugier, *Un ministère réformateur sous Louis XV*, is more recent and has useful details but is uncritical in the extreme. The most reliable brief recent account is in Egret, *Louis XV et l'opposition*, ch. V.
[35] Bosher, *French Finances*, 145-53. See also above, p.
[36] For this attitude, see Marion, *Histoire financière*, I, 259; Hudson, 'Parlementary Crisis', 101-2.
[37] Doyle, 'Parlements', 434-8; Egret, *Louis XV et l'opposition*, 220-3.
[38] e.g. Cobban, *History of Modern France* I, 101; Méthivier, *Siècle de Louis XV*, 125-6, and *La Fin de l'Ancien Régime*, 26-7.
[39] Doyle, 'Parlements', 441-50. See the same author's *Parlement of Bordeaux*, chs. XI-XVI, and Egret, *Le Parlement de Dauphiné*, II, chs. I-III. See, too, B. Stone, *The French Parlements and the Crisis of the Old Regime* (Chapel Hill, 1986), passim.
[40] Grange, *Idées de Necker*, 383-90; Egret, *Necker*, 131-2.

Notes to Chapter 4

[1] M. Fleury and A. Valmary, 'Les Progrès de l'instruction élémentaire de Louis XIV à Napoléon III d'après l'enquête de Louis Maggiolo (1877-1879)', *Population* xii (1957); see too F. Furet and W. Sachs, 'La Croissance de l'alphabétisation en France, XVIIIe-XIXe siècles', *Annales* xxix (1974), 714-37.
[2] R. Mandrou, *De la culture populaire aux 17e et 18e siècles* (Paris, 1964); G. Bollême, *Les Almanachs populaires au XVIIe et XVIIIe siècles. Essai d'histoire sociale* (Paris, 1969). The latter sees more developments in this literature over the eighteenth century than the former, but doubts about Bollême's conclusions can be found in R. Darnton, 'In Search of the Enlightenment: Recent Attempts To Create a Social History of Ideas', *J.M.H.* xliii (1971), 125-7.
[3] R. Estivals, *La Statistique bibliographique de la France sous la monarchie au XVIIIe siècle* (Paris/The Hague, 1965), 340, 346-7, 367, 380, 410-11.
[4] Mornet, *Origines intellectuelles*, 160.
[5] C. Bellanger, *et al. Histoire générale de la presse française* (Paris, 1969) I, 171.
[6] Mornet, *Origines intellectuelles*, 160.
[7] Ibid.
[8] R. Darnton, 'The Encyclopédie Wars of Prerevolutionary France',

A.H.R. lxxviii (1973), 134–6; see too, *The Business of Enlightenment.*

9 See D. Roche, 'Milieux académiques provinciaux et société des lumières', in F. Furet (ed.), *Livre et sociéte dans la France du XVIIIe siècle,* I (Paris, 1965), 94–7; *Le siècle des lumières en province* (2 vols., Paris, 1978).

10 Mornet, *Origines intellectuelles,* 306–7, 316–17.

11 D. Roche, 'Négoce et culture dans la France du XVIIIe siècle', *R.d'Hist. Mod. et Contemp.* xxv (1978), 380–4.

12 Mornet, *Origines intellectuelles,* 311.

13 J. Klaits, *Printed Propaganda under Louis XIV* (Princeton, N.J., 1976).

14 Grosclaude, *Malesherbes* 63–80, 101–86; F. Furet, 'La "Librairie" du royaume de France au 18e siècle' in Furet, *Livre et société, I, 4–14.*

15 Quoted in Grosclaude, *Malesherbes,* 165–6.

16 Darnton, 'Memoirs of Lenoir', 543.

17 R. Darnton, 'Le Livre français à la fin de l'Ancien Régime', *Annales* xxviii (1973), 743–4; 'Reading, Writing and Publishing in 18th Century France: A Case Study in the Sociology of Literature', *Daedalus* c (1971), 232–7.

18 See Egret, *Louis XV et l'opposition,* 230–1; Doyle, 'Parlements', 453–4.

19 R. Darnton, 'The Grub-Street Style of Revolution: J.-P. Brissot, Police Spy' *J.M.H.* xlvi (1968), esp. 320–6; see, too, R.C. Cobb, 'The Police, the Repressive Authorities and the beginning of the Revolutionary Crisis in Paris', *Welsh History Review* iii (1966–7), 430; Williams, *Police of Paris,* 104–11.

20 Darnton, 'Memoirs of Lenoir', 543.

21 D.C. Hudson, 'In Defense of Reform', 51–76.

22 See Doyle, *Parlement of Bordeaux,* 239–41.

23 Flammermont, *Le Chancelier Maupeou,* 435–7.

24 Doyle, 'Parlements', 438–9, 453–4.

25 L. Léouzon Le Duc, *Correspondance diplomatique du Baron de Staël-Holstein* (Paris, 1881), 32.

26 R. Darnton, 'The High Enlightenment and the Low Life of Literature', 54–65.

27 J. Erhard and J. Roger, 'Deux périodiques français du 18e siècle: "Le Journal des Savants" et "Les Mémoires de Trévoux". Essai d'une étude quantitative', in Furet, *Livre et société,* I, 33–57; H. Vyerberg, 'The Limits of Nonconformity in the Enlightenment. The Career of Simon Nicolas Henri Linguet', *Fr.Hist.St.* vi (1970), 474–91. D.G. Levey, *The Ideas and Careers of Simon Nicolas Henri Linguet* (Urbana, 1980).

28 J.F. Bosher, *The Single Duty Project: A study of the Movement for a French Customs Union in the eighteenth century* (London, 1962) ch. III; J.H. Langbein, *Torture and the Law of Proof. England and Europe under the Ancien Regime* (Chicago, 1977) chs. 3–4.

[29] J. Brancolini and M.-T. Bouyssy, "La Vie provinciale du livre à la fin de l'Ancien Régime' in F. Furet, *Livre et société* II (Paris, 1970), 9-15.

[30] Furet in *Livre et société*, I, 18-22.

[31] Taylor, 'Revolutionary and Nonrevolutionary Content', *Fr.Hist. St.* (1972).

[32] Darnton, 'The High Enlightenment and the Low Life of Literature', *passim.*

[33] Darnton, 'Encyclopédie Wars', 134-52; see too Darnton, 'Le Livre français à la fin de l'ancien régime', 739-40.

[34] See E. Carcassonne, *Montesquieu et le problème de la constitution française au XVIIIe siècle* (Paris, 1927), 262--96.

[35] D. Echeverria, 'The Pre-revolutionary Influence of Rousseau's *Contrat Social*', *Journal of the History of Ideas* xxxiii (1972), 543-60; N. Hampson, 'The *Recueil des pièces intéressantes*', 399-403.

[36] Furet in *Livre et société*, I.

[37] M. Vovelle, *Piété baroque et déchristianisation* (Paris, 1973).

[38] M. Agulhon, *Pénitents et Francs-Maçons de l'ancienne Provence* (Paris, 1968).

[39] P. Chevallier, *Histoire de la Franc-Maçonnerie en France;* I *La Maçonnerie: école de l'égalité 1725-1799* (Paris, 1974); Roberts, *Mythology of the Secret Societies.*

[40] Van Kley, *The Jansenists and the Expulsion of the Jesuits from France.*

[41] See, for example, the conclusions of B. Robert Kreiser, *Miracles, Convulsions and Ecclesiastical Politics in Early Eighteenth Century Paris* (Princeton, N.J., 1978), 395-401.

[42] P. Chevallier, *Loménie de Brienne et l'ordre monastique 1766-1789* (2 vols., Paris, 1959/60).

[43] D.D. Bien, *The Calas Affair. Persecution, Toleration and Heresy in 18th Century Toulouse* (Princeton, N.J., 1961).

[44] Mornet, *Origines intellectuelles*, 240-1.

[45] Ibid., 249; see, too, M. Marion, *Le Garde des Sceaux Lamoignon et la réforme judiciaire de 1788*, 33-7.

[46] P. Gay, *Voltaire's Politics. The Poet as Realist* (Princeton, N.J., 1959).

[47] See R. Derathé, 'Les Philosophes et le despotisme' in P. Francastel (ed.), *Utopie et institutions au XVIIIe siècle* (Paris, 1963), 57-75.

[48] See Palmer, *Age of the Democratic Revolution*, I, 60.

[49] See the examples in Darnton, 'The High Enlightenment and the Low Life of Literature', 79-83.

[50] Kaplan, *Bread, Politics and Political Economy*, I, ch. III.

[51] Ibid., II, chs. IX and X.

[52] Ibid., I, chs. IV-VII; II, chs. XI and XII.

[53] Ibid., I, 345-77 and 389-400, has convincingly traced the genesis and progress of this notorious idea.

[54] Hudson, 'In Defense of Reform', 72.

[55] F. Diaz, *Filosofia e politica nel settecento francese* (2nd edn., Turin, 1973), 444-70. On the exception of Voltaire, see Gay, *Voltaire's Politics*, 309-30; but also R.S. Tate, 'Voltaire and the "Parlements": A Reconsideration', *Studies on Voltaire and the Eighteenth Century* xc (1972).

[56] Carcassonne, *Montesquieu et le problème*, 401-67; Richet, 'Autour des origines', 20-1.

[57] K.M. Baker, 'French Political Thought at the Accession of Louis XVI', *J.M.H.* 1 (1978), 279-303, D. Echeverria, *The Maupeou Revolution. A Study in the History of Libertarianism: France, 1770-1774* (Baton Rouge, 1985).

[58] See Cobban, *Aspects*, 83-9.

[59] Mornet, *Origines intellectuelles*, 263-4; D. Jarrett, *The Begetters of Revolution. England's Involvement with France 1759-1789* (London, 1973), 72-9.

[60] R. Bickart, *Les Parlements et la notion de souveraineté nationale au XVIIIe siècle* (Paris, 1932).

[61] R.A. Leigh in *Historical Journal* (1969), 561-4.

[62] Echeverria, 'The Pre-revolutionary Influence of Rousseau's *Contrat Social*'.

[63] *Mémoire sur les états provinciaux* (1750).

[64] Confusingly entitled *Mémoire sur les municipalités*. See Dakin, *Turgot*, 272-80, and Baker, 'French Political Thought', 295-8.

[65] J. de Witte (ed.), *Journal de l'Abbé de Véri* (Paris, 1933), II, 8.

[66] Egret, *Parlement de Dauphiné*, II, 125-40; Doyle, *Parlement of Bordeaux*, 227-8.

[67] See G. Boscary, *L'Assemblée provinciale de Haute-Guyenne 1779-1790* (Paris, 1932); also Grange, *Idées de Necker*, 366-90.

[68] F. Acomb, *Anglophobia in France 1763-89* (Durham, N.C., 1950); G. Bonno, *La Constitution britannique devant l'opinion française de Montesquieu à Bonaparte* (Paris, 1932).

[69] Palmer, *Age of the Democratic Revolution*, I, ch. IX; and in more detail D. Echeverria, *Mirage in the West: A History of the French Image of American Society to 1815* (Princeton, N.J., 1956).

Notes to Chapter 5

[1] The most accessible outlines are Goodwin, 'Calonne', 210-34, and Egret, *Pré-Révolution*, 7-9.

[2] Bosher, *The Single Duty Project*, ch. XI.

[3] See Doyle, *Parlement of Bordeaux*, ch. XV.

[4] Quoted in Egret, *Pré-Révolution*, 9.

[5] Ibid. The most important printed source on the Notables is P. Chevallier (ed.), *Journal de l'Assemblée des Notables de 1787* (Paris, 1960), being the notes of the Brienne brothers; see too V. Gruder, 'Class and Politics in the Pre-Revolution: The Assembly of Notables of 1787' in E. Hinrichs, E. Schmitt, and R. Vierhaus, *Vom Ancien Régime zur Französischen Revolution* (Göttingen, 1978) 207-32.

6 Egret, *Pré-Révolution,* 13.
7 Quoted ibid., 8.
8 Ibid., 25.
9 Goodwin, 'Calonne', 360–1.
10 Quoted ibid., 373 n. 3.
11 Quoted in Egret, *Pré-Révolution*, 60.
12 As does Goodwin, 'Calonne', 373–7; Gruder, 'Class and Politics', prefers to see the resistance as that of a class of landowners.
13 For a survey of these reforms, see Egret, *Pré-Révolution,* 61–9, 73–146.
14 See Bosher, *French Finances,* 196–214.
15 Quoted in Egret, *Pré-Révolution,* 55.
16 Doyle, *Parlement of Bordeaux,* 265–75.
17 Quoted in Egret, *Pré-Révolution,* 168.
18 O. Browning (ed.), *Despatches from Paris 1784–1790* (Camden Third Series, vol. XVI, London, 1909), I, 212.
19 Egret, *Pré-Révolution,* 190. The British ambassador heard this too; see Browning, *Despatches,* 267.
20 Quoted in Egret, *Pré-Révolution,* 191.
21 As modified by Brienne. Originally Calonne had planned a proportional tax.
22 The fullest recent analysis of the reforms is in Egret, *Pré-Révolution,* 122–33, 246–305.
23 See ibid., 267–78; Hampson, 'The *Recueil des pièces intéressantes'*.
24 Egret, *Pré-Révolution,* 279–90; Dawson, *Provincial Magistrates and Revolutionary Politics,* 137–49.
25 For this 'Noble Revolt', see below, p. 141.
26 See below, p. 142.
27 Egret, *Pré-Révolution,* 263–5; Pillorget 'Les Problèmes du maintien de l'ordre', 118–19; S.F. Scott, *The Response of the Royal Army to the French Revolution* (Oxford, 1978), 46–8.
28 See below, ch. 9.
29 Egret, *Pré-Révolution,* 311–15.
30 Bosher, *French Finances,* 198–9.

Notes to Chapter 6

1 The most convenient guide to the literature on these questions is Doyle, 'Was There an Aristocratic Reaction?', 13–20.
2 Meyer, 'La Noblesse Française', 44.
3 Ibid., 28–37.
4 Richard, *Noblesse d'affaires.*
5 Durand, *Les Fermiers-Généraux au XVIIIe siècle,* 288–301.
6 Chaussinand-Nogaret, *La Noblesse au XVIIIe siècle,* 150–8.
7 See Behrens, *The Ancien Régime,* 46–7, 58–62.
8 e.g. in 1776, in the remonstrances of the parlement of Paris against the commutation of the *corvée.* J. Flammermont, *Les Remonstrances du parlement de Paris au XVIIIe siècle* (3 vols., Paris,

1888-98), III. See, too, O. Niccoli, *I sacerdoti, i guerrieri, i contadini. Storia di un' immagine della società* (Turin, 1979).

9 See the debate between B. Behrens and G.J. Cavanaugh in *Fr. Hist. St.* (1974), 681-92, and (1975).

10 Richard, *Noblesse d'affaires*, 21-52.

11 Doyle, 'Was There an Aristocratic Reaction?', 17-19; Bien, 'La Réaction aristocratique avant 1789', *passim*; Meyer, 'La Noblesse française', 22-7.

12 Ford, *Robe and Sword*, ch. IV.

13 The lower figure is Chaussinand-Nogaret's, *La Noblesse au XVIIIe siècle*, 40-6; the higher is Meyer's, 'La Noblesse française', 8-17.

14 For an exploration of some of these issues, see J.Q.C. Mackrell, *The Attack on 'Feudalism' in Eighteenth Century France* (London, 1973).

15 *L'Ami des Hommes* (1756).

16 *De l'Esprit des Lois* (1748), Bk. II, ch. IV.

17 C. Van Doren, *Benjamin Franklin* (New York, 1938), 710.

18 See W. Doyle, 'Price of Offices', *passim.*

19 Doyle, 'Parlements', 444.

20 Ford, *Robe and Sword*, 246-52 and *passim.*

21 Stone, 'Robe against Sword'.

22 P. du Puy de Clinchamps, *La Noblesse* (Paris, 1959), 64.

23 What follows is largely based upon Chaussinand-Nogaret, *La Noblesse au XVIIIe siècle*, 77-8.

24 R. Forster, *The House of Saulx-Tavanes. Versailles and Burgundy 1700-1830* (Baltimore, Mal., 1971), 55-6, shows that a third of this not untypical courtier family's income in the 1780s came from pensions and sinecures. Ch. III shows vividly how such revenues were spent.

25 Bien, 'Réaction aristocratique', 41-2, 515-16.

26 Ibid., 515-19.

27 Chaussinand-Nogaret, *La Noblesse au XVIIIe siècle*, Ch. IV.

Notes to Chapter 7

1 See Vovelle and Roche, 'Bourgeois, Rentiers and Property Owners', in Kaplow, *New Perspectives*, 25-46.

2 Quoted in G. Lewis, *Life in Revolutionary France* (London, 1972), 76.

3 Cobban, *Social Interpretation*, 107-9, 117-19, conducts a devastating analysis of this concept.

4 P. Léon in Braudel and Labrousse, *Histoire économique et sociale*, 607. Since his estimate of the nobility's numbers is rather higher than any other recent suggestion, these figures should probably be revised upwards rather than downwards.

5 Taylor, 'Noncapitalist Wealth', 486.

6 Labrousse in *Histoire économique et sociale*, 476.

7 See Barber, *Bourgeoisie*, ch. IV and *passim.*

8 Meyer, *La Noblesse bretonne au XVIIIe siècle*, I, 257-60; Doyle,

'Le Prix des charges anoblissantes à Bordeaux au XVIIIe siècle', *Annales du Midi* lxxx (1968).

9 Hufton, *Bayeux*, 57-8; Garden, *Lyon et les Lyonnais*, 356. I arrive at this figure by deducting artisans and nobles from the 4,000 rich taxpayers of 1790.

10 Sentou, *Fortunes et groupes sociaux à Toulouse sous la Révolution* 146; P. Butel, *Les Négoçiants bordelais. L'Europe et les Iles au XVIIIe siècle* (Paris, 1974), 281-5; Doyle, *Parlement of Bordeaux* 16-17, 53-5.

11 Butel, *Négoçiants*, 285; Sentou, *Fortunes*, 151.

12 D. Ligou, *Montauban à la fin de l'Ancien Régime et aux débuts de la Révolution 1787-94* (Paris, 1958), 70; J.N. Hood, 'Protestant -Catholic Relations and the Roots of the First Popular Counter-Revolutionary Movement in France', *J.M.H.* xliii, (1971); and 'Revival and Mutation of Old Rivalries in Revolutionary France', *P. and P.* 82 (1979).

13 Lucas, 'Nobles, Bourgeois, etc', 97.

14 Within the nobility, of course, the matter was not quite so straightforward. See above, p. 118.

15 Lucas, 'Nobles, Bourgeois, etc.', 114.

16 Labrousse in *Histoire économique et sociale*, 386-91, 396-406, 454-63, 477.

17 L. Tuetey, *Les Officiers sous l'Ancien Régime. Nobles et roturiers* (Paris, 1908), 218.

18 J. Tarrade, *Le Commerce colonial de la France à la fin de l'Ancien Régime: l'évolution du régime de l'Exclusif de 1763 à 1789* (2 vols., Paris, 1972); Butel, *Négoçiants*, 38-9.

19 They were certainly serious in Lyons: Garden, *Lyon et les Lyonnais*, 178, 277.

20 M. Bouloiseau, *Cahiers de doléances du tiers état du bailliage de Rouen* (2 vols., Paris, 1957).

21 Taylor, 'Noncapitalist Wealth', *passim*.

22 Hufton, *Bayeux*, 62; Berlanstein, *Barristers*, 44; Gresset, *Monde judiciaire*, I, 490.

23 In addition to those cited in the previous footnote, see Dawson, *Provincial Magistrates*, ch. III.

24 Hufton, *Bayeux*, 62-3; Berlanstein, *Barristers*, 11-23; Gresset, *Monde judiciaire*, I, 280-92.

25 R.L. Kagan, 'Law Students and Legal Careers in Eighteenth-Century France', *P and P* 68 (1975).

26 Doyle, 'Price of Offices', *passim*.

27 Lucas, 'Nobles, Bourgeois, etc.', 119-20.

28 Dawson, *Provincial Magistrates*, 60-65, 129-49.

29 F. Delbeke, *L'Action politique et sociale des avocats au XVIIIe siècle* (Paris/Louvain, 1927), 90-108; Doyle, *Parlement of Bordeaux*, 160-1, 189; M. Gresset, *Le Monde judiciaire à Besançon (1647-1789)* (Lille, 1975), II, 1161-95; Egret, *Parlement de Dauphiné*, II, 85-9.

30 Barber, *Bourgeoisie*, 144-5.

31 Taylor, 'Revolutionary and Nonrevolutionary Content', *passim*; Roche, 'Négoce et culture dans la France du XVIIIe siècle', *passim*.

32 Richet, *La France moderne*, 152; 'Autour des origines lointaines'.

Notes to Chapter 8

1 Egret, *Pré-Révolution*, 315-6; Egret, *Necker*, 217-22, 226-9; Chapuisat, *Necker*, 146-54.

2 It was too late to prevent the meetings in Dauphiné. See below p. 142.

3 So argued Malesherbes in July; see Egret, *Pré-Révolution*, 322.

4 Hailes to Carmarthen, 9 Oct. 1788, Browning, *Despatches*, II, 107.

5 Egret, *Necker*, 234-5; B.S. Stone, 'Crisis in the Paris Parlement: The Grand' Chambriers, 1774-1789' (Ph.D. dissertation, Princeton, N.J., 1972), 296; Doyle, *Parlement of Bordeaux*, 292-3.

6 Stone, 'Crisis', 151-208.

7 Some 'patriotic' pamphleteers shared the parlements' suspicions, e.g. Billaud-Varenne, *Du despotisme des ministres de France* (Amsterdam, 1789), 3 vols., II, 371: 'Si, ne pouvant plus reculer, l'administration essayait de former une convocation des états suivant les principes qu'elle manifeste, sans doute cette nouvelle opération, illégale par sa nature, ne pourrait jamais produire que des effets nuls.'

8 Egret, *Pré-Révolution*, 116-17; see too his 'La Pré-Révolution en Provence, 1787-88, *A.H.R.F.* xxvi (1954), trans. in Kaplow, *New Perspectives*, 153-69.

9 Egret, *Parlement de Dauphiné*, II, 125-40, 169-296.

10 Lefebvre, *Coming*, 44-7.

11 Eisenstein, 'Who Intervened in 1788?'; G. Michon, *Essai sur l'histoire du parti Feuillant. Adrien Duport* (Paris, 1924), 24-39; Egret, *Pré-Révolution*, 326-30.

12 'conspiration d'honnêtes gens'; Michon, *Duport*, 31. D.L. Wick, 'The Court Nobility and the French Revolution. The Example of the Society of Thirty', *Eighteenth-Century Studies* xiii (1980).

13 See Egret, *Pré-Révolution*, 339-46; *Necker*, 236-40.

14 Egret, *Pré-Révolution*, 331-9.

15 Ibid., 354-8.

16 Northern and southern terms, respectively, for the same judicial unit.

17 Egret, *Pré-Révolution*, 347-51; Stone 'Crisis', 301-6.

18 My translation. For another, see Lefebvre, *Coming*, 45.

19 Doyle, *Parlement of Bordeaux*, 289-92.

20 Egret, 'La Pré-Révolution en Provence'; Kaplow, *New Perspectives*, 155-7, 160-1.

21 J. Egret, 'La Révolution aristocratique en Franche Comté et son échec (1788-1789)', *Rev. d'hist.mod. et contemp.* ii (1954); C. Brelot, *La Noblesse en Franche Comté de 1789 à 1808 (Paris, 1972), 49-52.*

[22] N. Hampson, 'The Enlightenment and the Language of the French Nobility in 1789: The Case of Arras' in D.J. Mossop, G.E. Rodmell, and D.B. Wilson (eds.), *Studies in the French Eighteenth Century Presented to John Lough* (Durham, 1978), 81-9.

[23] J. Egret, 'Les Origines de la Révolution en Bretagne (1788-1789)', *Revue historique* ccxxi (1955), trans. in Kaplow, *New Perspectives,* 136-52.

[24] See B.F. Hyslop, *A Guide to the General Cahiers of 1789* (2nd edn., New York, 1968), 11-31; Egret, *Necker,* 248-67.

[25] J.M. Roberts, *The French Revolution* (Oxford, 1978), 93.

[26] Quoted in J. McManners, *The French Revolution and the Church* (London, 1969), 18. See too M.G. Hutt, 'The Curés and the Third Estate: The Ideas of Reform in the pamphlets of the French lower Clergy in the Period 1787-1789', *Journal of Ecclesiastical History* viii (1957).

[27] R.F. Necheles, 'The Curés in the Estates General of 1789', *J.M.H.* xlvi (1974), 427.

[28] e.g. at Bayeux, Hufton, *Bayeux,* 135-8; or at Angers, J. McManners, *French Ecclesiastical Society under the Ancien Régime. A Study of Angers in the Eighteenth Century* (Manchester, 1960), 220-9.

[29] Quoted in E. Champion, *La France d'après les cahiers de 1789* (Paris, 1897), 177.

[30] J. Murphy and P. Higonnet, 'Les Députés de la noblesse aux États-Généraux de 1789', *R.d'Hist. Mod. et Contemp.* xx (1973), 238-9.

[31] Brelot, *Noblesse en Franche Comté,* 54-5.

[32] Murphy and Higonnet; H. Carré, *La Fin des parlements (1788-90)* (Paris, 1912), 94-9.

[33] Murphy and Higonnet, 240-3.

[34] Chaussinand-Nogaret, *Noblesse au XVIIIe siècle,* 183-226.

[35] e.g. at Bordeaux there were clashes between town and country delegates. See A. Forrest, *Society and Politics in Revolutionary Bordeaux* (Oxford, 1975), 34-5.

[36] Cobban, 'Myth of the French Revolution', reprinted in *Aspects,* 110-11.

[37] cf. Cobban, *Social Interpretation,* 59-61.

[38] Dawson, *Provincial Magistrates,* 186-92.

[39] Taylor, 'Revolutionary and Nonrevolutionary Content', 500.

[40] Michon, *Duport,* 35, 38; Taylor, 'Revolutionary and Nonrevolutionary Content', 496.

[41] Taylor, 'Revolutionary and Nonrevolutionary Content', 497-9.

Notes to Chapter 9

[1] Bosher, *French Finances,* 183-91, 198-9.

[2] Morineau, *Les Faux semblants, passim,* and Dupâquier, *La Population française, passim.*

[3] Bosher, 'French Crisis of 1770', *passim.*

[4] Quoted in Braudel and Labrousse, *Histoire économique et sociale de la France,* II, 548-9.

[5] Labrousse, *La Crise*, II, 597-608.
[6] Rudé, *Crowd in the French Revolution*, 31-3.
[7] Dorset to Carmarthen, 8 Jan. 1789, Browning, *Despatches*, II, 140.
[8] F. Crouzet, 'England and France in the 18th Century: A Comparative Analysis of Two Economic Growths', in R.M. Hartwell (ed.), *The Causes of the Industrial Revolution in England* (London, 1967), 146-7.
[9] J. Meyer, *L'Armement nantais dans la deuxième moitié du XVIIIe siècle* (Paris, 1969); P. Butel, *Négoçiants, passim.*
[10] Champion, *La France d'après les cahiers*, 163-4.
[11] Kaplan, *Bread, Politics and Political Economy, passim.*
[12] Rudé, *Crowd in History*, 23-31.
[13] See J. Godechot, *The Taking of the Bastille, July 14th 1789* (London, 1970), 128-33.
[14] See Rudé, *Crowd in the French Revolution*, 34-44; Godechot, *Bastille*, 136-51.
[15] Scott, *Response of the Royal Army*, 48-51; Godechot, *Bastille*, 131-2.

Notes to Chapter 10
[1] Murphy and Higonnet, 233.
[2] Ibid., 235-7.
[3] e.g. Creuzé-Latouche, magistrate and deputy from Chatellerault, recorded on 19 June the opinion of a conservative noble that the third estate was overrepresented. 'This', he noted, 'is in the mouths of all nobles and in their hearts', and he went on to denounce the 'perfidy with which this Order abuses us when hiding behind the interests of the monarchy to retain the power of oppressing us. History offers us no example of an Order in which the *esprit de corps* has been so high, so anti-social and so animated against the public good.' J.A. Creuzé-Latouche, *Journal des États-Généraux, 18 Mai–29 Juillet 1789*, J. Marchand (ed.), (Paris, 1946), 129-30.
[4] Ibid., 12-22.
[5] Scott, *Response*, 51-2; Egret, *Necker*, 289; G. Lefebvre, *The Great Fear of 1789* (Eng. trans. London, 1973), 59-61.
[6] M.G. Hutt, 'The Role of the Curés in the Estates-General', *Journal of Ecclesiastical History* vi (1955); Necheles, 'The Curés in the Estates-General'.
[7] The original date fixed was the 22nd, but subsequently there was a day's postponement.
[8] Egret, *Necker*, 288-98.
[9] C. Maxwell (ed.), *Travels in France during the Years 1787, 1788 and 1789* (Cambridge, 1929), 159.

Notes to Chapter 11
[1] *Travels*, 176.
[2] Daumard and Furet, *Structures et relations sociales à Paris*; see too Godechot, *Bastille*, 47-50.

[3] J. Kaplow, *The Names of Kings. The Parisian Laboring Poor in the Eighteenth Century* (New York, 1972), 16.
[4] See O. Hufton, 'Towards an Understanding of the Poor of Eighteenth Century France', in J.F. Bosher (ed.), *French Government and Society 1500–1850* (London, 1973).
[5] G. Rudé, 'Cities and Popular Revolt 1750-1850', Ibid.
[6] S. Kaplan, 'Réflexions sur la police du monde de travail, 1700–1815', *Revue historique* cclix (1979), 30.
[7] Ibid., *passim.*
[8] Kaplow, *Names*, 38–9; Rudé, *Crowd in the French Revolution*, 20–1.
[9] H. Simpson (ed.), *The Waiting City. Paris 1782–88* (London, 1933), 108–9.
[10] Williams, *Police of Paris*, 67–111; Godechot, *Bastille*, 78–86.
[11] Kaplow, *Names*, 154–8; its acquiescence in August 1787 in Brienne's edict abandoning controls seems to have gone unnoticed.
[12] Egret, *Pré-Révolution* 318–19.
[13] *Travels*, 134–5.
[14] Browning, *Despatches*, II, 226.
[15] See Scott, *Response*, 54–9.
[16] Rudé, *Crowd in the French Revolution*, 48–9.
[17] Browning, *Despatches*, II, 240.

Notes to Chapter 12

[1] *Travels*, 173.
[2] R. Forster, 'Obstacles to Agricultural Growth in eighteenth century France', *A.H.R.* lxxv (1970), 1604.
[3] Hufton, *The Poor of Eighteenth Century France*, chs. II and III *passim.*
[4] See Cobban, *Social Interpretation* 110–18; but compare Doyle, *Parlement of Bordeaux*, 73, and A. Davies, 'The Origins of the French Peasant Revolution of 1789', *History* xlix (1964), 32–4.
[5] Hincker, *Les Français devant l'impôt*, 48.
[6] The fullest recent survey of the workings of feudalism is J. Bastier, *La Féodalité au siècle des lumières dans la région de Toulouse (1730–1790)* (Paris, 1975).
[7] Ibid., 262–4.
[8] Ibid., 260.
[9] See W. Doyle, 'Was There an Aristocratic Reaction?', 114–20; also M. Garaud, 'Le Régime féodal en France à la veille de son abolition' in J. Godechot (ed.), *L'Abolition de la féodalité dans le monde occidental* (2 vols., Paris, 1971) I, 111–12.
[10] Bastier, 242–57, takes its existence for granted, but nothing in his very full and careful account of how terriers were revised proves that this was being done with renewed vigour in the late eighteenth century. Many of his examples come from before 1750.
[11] Cobban, *Social Interpretation*, 47–52.
[12] Taylor, 'Revolutionary and nonrevolutionary Content', 495–6.

[13] Mackrell, *The Attack on 'Feudalism' in eighteenth century France.*

[14] Godechot, *Bastille*, ch. VII.

[15] G. Lefebvre, *The Great Fear of 1789* (1932, Eng. trans. 1973), 38–40.

[16] Quoted ibid., (1973), 40.

[17] M. Marion, 'Le Recouvrement des impôts en 1790', *Revue Historique* (1916); see too R.B. Rose, 'Tax Revolt and Popular Organisation in Picardy, 1789-91', *P. and P.* 43 (1969).

[18] Lefebvre, *Great Fear*, 101 ff.

[19] Lefebvre is emphatic that the revolts and the fear were quite separate phenomena. Ibid., 120–1, 142.

[20] Taylor 'Revolutionary and nonrevolutionary Content', 498.

[21] P. Kessel, *La Nuit du 4 août 1789* (Paris, 1969).

[22] H. Carré (ed.), *Marquis de Ferrières. Correspondance inédite 1789, 1790, 1791* (Paris, 1932) 114–15.

Notes to Chapter 13

[1] For this reason Lefebvre ends his *Coming of the French Revolution* with the October Days; but H. Méthivier, *La Fin de l'Ancien Régime* (Paris, 1970), ends in August.

[2] Taylor, 'Revolutionary and nonrevolutionary Content', 485–9.

[3] See Cobban, *Social Interpretation*, 25–53, and Mackrell, *The Attack on 'Feudalism' in eighteenth century France*, 174–7.

[4] Taylor, 'Revolutionary and nonrevolutionary Content', 499.

[5] Ibid., 498.

[6] Ibid., Chaussinand-Nogaret, *La Noblesse au XVIIIe siècle*, 213.

[7] Dawson, *Provincial Magistrates*, 223–5. For the inadequacy of the compensation actually voted, see Doyle, 'Price of Offices' 848, 859–60.

[8] Chaussinand-Nogaret, *La Noblesse au XVIIIe siècle*, 210, 222.

[9] P.V. Malouet, *Mémoires* (Paris, 1874), I., 265–7.

Index of Authors Cited

General Index